MATHEMATICS RESEARCH DEVELOPMENTS

A NOTE ON INHERITANCE AND GENERALIZABILITY PROPERTIES IN OPTIMAL CONTROL PROBLEMS

MATHEMATICS RESEARCH DEVELOPMENTS

Additional books in this series can be found on Nova's website under the Series tab.

Additional e-books in this series can be found on Nova's website under the eBooks tab.

MATHEMATICS RESEARCH DEVELOPMENTS

A NOTE ON INHERITANCE AND GENERALIZABILITY PROPERTIES IN OPTIMAL CONTROL PROBLEMS

IRAJ SADEGH AMIRI
EDITOR

NOVINKA
New York

Library of Congress Cataloging-in-Publication Data

Names: Amiri, Iraj Sadegh, 1977- editor.
Title: A note on inheritance and generalizability properties in optimal control problems / Iraj Sadegh Amiri (senior lecturer, Photonics Research Centre (PRC), University of Malaya (UM), Pantai Hillpark, Pantai Murni, Kuala Lumpur, Malaysia), editor.
Description: Hauppauge, New York : Nova Science Publishers, Inc., [2016] | Series: Mathematics research developments | Includes index.
Identifiers: LCCN 2016029815 (print) | LCCN 2016039719 (ebook) | ISBN 9781634857840 (softcover) | ISBN 9781634857963
Subjects: LCSH: Convex functions. | Subdifferentials.
Classification: LCC QA331.5 .N68 2016 (print) | LCC QA331.5 (ebook) | DDC 515/.724--dc23
LC record available at https://lccn.loc.gov/2016029815

Published by Nova Science Publishers, Inc. † New York

CONTENTS

PREFACE

This research consists of two parts. The first part focuses on the development of the proposition that "if there exists a type of function, then there exists a functional with the same type" based on the proposition of the Inheritance and Generalizability properties of the function in a functional. This study presents the abstract convex, increasing positively homogeneous and convex-along-rays functionals via this proposition. The second part is on the investigation of the global search optimization algorithm use called the Cutting Angle Method (CAM) on Optimal Control Problems (OCP). Many algorithms are available for solving OCP, but they are basically local search algorithms. To overcome the problem associated with local searches, most OCP are modelled as Linear Quadratic Regulator (LQR) problems in the hope that the solution found estimates the true global solution of the original problem. However, in doing so, a lot of information carried by the original problem might be lost in its translation into LQR models. CAM being a global search algorithm is expected to overcome this problem. It can be used alone or in combination with a local search to find the global solution. CAM has been successfully used on functions; however, OCPs are functionals. To do this, a model has been introduced based on Inheritance and Generalizability properties to demonstrate that the optimization algorithms which are used for functions can also be extended for use in functionals. Based on these properties, the study discovered that with the Unit Vectors Combinations Technique (UVCT) proposed in this research, and CAM could successfully work on functionals in general and with OCP particularly. To help speed up the convergence of CAM, the literature proposed the use of local searches for the determination of the initial solution. In a case study done in the research, CAM was successfully combined with a local search named the Dynamic

Integrated System Optimization and Parameter Estimation (DISOPE) algorithm. Moreover, the initial solution given by DISOPE algorithm has been verified as global by CAM.

Chapter 1 – Convexity as a main field of Euclidean geometry has its proof, in the language of algebra and analysis. This study proposes abstract convex functionals and investiagtes the existence of their subdifferentiability which is defined on $\mathbb{R}^n$ based on the Inheritance and Generalizability of function properties on functionals.

Chapter 2 – As a major branch of Euclidean geometry, convexity has its proof in the field of algebra and analysis. This study investigates convex-along-rays functionals and based on the Inheritance and Generalizability of function properties on functionals it analyses the existence of their subdifferentiability which is defined on $\mathbb{R}^n$.

Chapter 3 – The homogeneous functionals which are defined on the cone $\mathbb{R}^n$ are well presented and developed. This paper discusses increasing positively homogeneous (IPH) functionals and based on the Inheritance and Generalizability of function properties on functionals, it investigates the existence of their sub differentiability on $\mathbb{R}^n$.

Chapter 4 – Optimal control is a study on way of choosing the best set of decisions to accomplish a particular objective. The paper mainly concerns the combinations of the Cutting Angle Method (CAM) with a local search in global optimization for solving functionals. Based on using the unit vector, a new technique is proposed to employ the CAM directly on optimal control problems. The numerical findings indicated that the proposed approach is very beneficial in the search for a global minimum.

Chapter 5 – Calculus of variations which is a field of mathematics deals with functionals. This study puts forward an application of the cutting angle method on a class of problems referred to as the variational problems. The solution approach is based on a recent technique which is called the unit vectors combinations. At the end of the study the numerical findings will be presented.

Chapter 6 – A study of optimal control is a way of taking a best set of decisions to accomplish a particular objective which deals with functionals. This paper focuses on an application of the cutting angle method to a class of problems referred to as the optimal control problems. The solution approach is based on a new technique called unit vector combinations. At the end of this study the numerical findings are put forward.

In: A Note on Inheritance …
Editor: Iraj Sadegh Amiri

ISBN: 978-1-63485-784-0

Chapter 1

ABSTRACT CONVEX FUNCTIONALS DEFINED ON $\mathbb{R}^n$ AND THEIR SUBDIFFERENTIABILITY

***Seyedalireza Seyedi*[1,*], *Iraj Sadegh Amiri*[2,†] *and Sara Chaghervand*[3,‡]**

[1]Department of Mathematics, Hamedan Branch,
Islamic Azad University, Hamedan, Iran
[2]Photonics Research Centre, University of Malaya,
Kuala Lumpur, Malaysia
[3]Department of Electrical Engineering, Borujerd Branch,
Islamic Azad University, Borujerd, Iran

ABSTRACT

Convexity as a main field of Euclidean geometry has its proof, in the language of algebra and analysis. This study proposes abstract convex functionals and investiagtes the existence of their subdifferentiability which is defined on $\mathbb{R}^n$ based on the Inheritance and Generalizability of function properties on functionals.

* seyedalireza.seyedi@gmail.com.
† Corresponding Author address. E-mail: amiri@um.edu.my.
‡ engsara1@yahoo.com.

Keywords: abstract convexity, subdifferential, functional

1. INTRODUCTION

Functional analysis, as a branch of mathematics, deals with vector spaces which have infinite dimensions as well as mappings existing between them. These mappings are called functionals on condition that the range is on the real line [1]. To analyse the properties of functionals based on the framework of functions, first some fundamental information about functions shall be presented. This information is collected from abstract analysis and topology and is presented as some definitions and theorems which are considered essential to be used in the current paper. Based on these, the study shall investigate the Inheritance and Generalizability of function properties for functionals. Then the study proposes Increasing Positively Homogeneous (IPH) functional on $\mathbb{R}^n$ based on function framework and those function properties.

A functional is traditionally a map which usually ranges from a vector space of functions to real numbers. To clarify this, usually a function takes functions as an argument or input and then returns a real number [2], [3]. The following examples show how Theorem 2.1 is proved through this paper.

Definition 1.1: A set is said to be finite if there is a bijective correspondence of $\mathcal{A}$ with some segment of the positive integers. That is, $\mathcal{A}$ is finite if it is empty or if there is a bijection $f: \mathcal{A} \to \{1, \ldots, n\}$ [4].

Definition 1.2: A set $\mathcal{A}$ is said to be countably infinite if there is a bijective correspondence $f: \mathcal{A} \to \mathbb{Z}_+$, [4].

Let $\mathcal{A}$ be a set of all functions defined on $\mathbb{R}^n$. A map is described from the set of natural numbers to set $\mathcal{A}$ which implies that set $\mathcal{A}$ is countable. The following theorems and corollary are crucial to the discussion.

Theorem 1.1: Let $\mathcal{A}$ be a nonempty set. Then the following are equivalent [4]:

- *A is countable;*
- *There is a surjective function* $\boldsymbol{f}: \mathbb{Z}_+ \longrightarrow \boldsymbol{\mathcal{A}}$*;*
- *There is an injective function* $\boldsymbol{g}: \boldsymbol{\mathcal{A}} \longrightarrow \mathbb{Z}_+$.

Theorem 1.2: A countable union of countable sets is countable [4].

Corollary 1.1: A subset of a countable set is countable [4].

Definition 1.3 (Indexed Family): Let $\mathcal{F}$ be a nonempty collection of sets. An indexing function for $\mathcal{F}$ is a surjective function f from set $\mathcal{E}$, called the index set, to $\mathcal{F}$. The collection $\mathcal{F}$, together with the indexing function f, is called an indexed family of sets. Given $d \in \mathcal{E}$, the set $f(d)$ shall be denoted by the symbol f_d. And the indexed family itself shall be denoted by the symbol$\{f_d\}_{d\in\mathcal{E}}$. Suppose that $f: \mathcal{E} \longrightarrow \mathcal{F}$ is an indexing function for $\mathcal{F}$; Let f_d denote $f(d)$, [4]. Then,

$$\bigcup_{d\in\mathcal{J}} f_d = \{x|\text{for at least one } d \in \mathcal{J}, x \in f_d\}.$$

Definition 1.4 (Axiom of Induction): The basic assumption of induction is, in logical symbols,

$$\forall \text{ predicates } \Theta, (\Theta(0) \wedge \forall k \in \mathbb{N}[\Theta(k) \Rightarrow \Theta(k+1)]) \Rightarrow \forall n \in \mathbb{N}\, \Theta(n)$$

where Θ is the proposition in question and k in addition to n are both natural numbers. In other words, the basis $\Theta(0)$ being true along with the inductive case ("$\Theta(k)$ is true implies $\Theta(k+1)$ is true for all natural k") being true together imply that $\Theta(n)$ is true for any natural number n [5].

2. Generalizability of Function Property for Functionals on $\mathbb{R}^n$

As it is understood from Definitions 1.3 and 1.4, in order to investigate the existence of the Inheritance and Generalizability properties, the procedure to construct a model shall be put forward. This model conventionally is called the Seyedi-Rohanin model (SRM):

Procedure 2.1 (SRM):

Categorize set $\mathcal{A}$ into all their subset groups via their names, i.e., let $\boldsymbol{\mathcal{B}_1}$ be a set of all convex functions on $\mathbb{R}^n$, $\boldsymbol{\mathcal{B}_2}$ be a set of all linear functions on

$\mathbb{R}^n$, etc. Then index all these subset groups via indexed family (see Definition 1.3) i.e., $\boldsymbol{f}_{d_1}$ for the set $\boldsymbol{\mathcal{B}}_1$, $\boldsymbol{f}_{d_2}$ for the set $\boldsymbol{\mathcal{B}}_2$, etc. Then,

$$\alpha: f(d) = f_d,$$

$$\alpha: \mathbb{Z}_+ \longrightarrow \mathcal{A}$$

Then,

$d \neq c \rightarrow f_d \neq f_c \rightarrow f(d) \neq f(c)$, implies α is injective and corange(α)= $\bigcup_1^\infty \mathcal{B}_m$= $\mathcal{A}$ implies that α are surjective, $\forall f_d \in \mathcal{A} \Rightarrow f_d = f(d), \exists\, d \in \mathbb{N}$. Thus, α is bijective.

The sets $\mathcal{B}_m$ are countably infinite ($m = 1,2, \ldots$). $\mathcal{A} = \bigcup_1^\infty \mathcal{B}_m$ is countably infinite, because of $\bigcup_1^\infty \mathcal{B}_m$ is the union of all countably infinite sets (see Theorem 1.2). It implies that ℕ and set $\mathcal{A}$ have a one-to-one correspondence. Thus, it implies that set $\mathcal{A}$ is countably infinite (see Theorem 1.1).

Let $\boldsymbol{\mathcal{Z}}$ be a set of all functionals defined on $\mathbb{R}^n$. Similarly, set $\boldsymbol{\mathcal{Z}}$ is categorized into all their subsets groups via their names, i.e., convex functionals, linear functionals, etc. Then all these groups are indexed.

Describe a map from set $\boldsymbol{\mathcal{A}}$ to set $\boldsymbol{\mathcal{Z}}$ to demonstrate their one-to-one correspondence. The following theorem shall establish the bijectivity of the map which was described between the set $\boldsymbol{\mathcal{A}}$ and set $\boldsymbol{\mathcal{Z}}$ in the model.

Theorem 2.1: Let the set $\mathcal{A}$ be the set of all functions defined on $\mathbb{R}^n$ and set $\mathcal{Z}$ be the set of all functionals defined on $\mathbb{R}^n$. If (α, β) represents the bijective maps from the natural number to both sets, then ψ as a map from set $\mathcal{A}$ to $\mathcal{Z}$is bijective.

Proof: By the assumption that α and β are bijective implies that

$$\begin{cases} \alpha \text{ is injective} \\ \beta \text{ is injective} \end{cases} \Rightarrow$$

$$\begin{cases} i \neq e \rightarrow f_i \neq f_e \rightarrow: \mathrm{f}(i) \neq: f(e) \\ i \neq e \rightarrow g_i \neq g_e \rightarrow: g(i) \neq: g(e) \end{cases} \tag{2.1}$$

$$\text{and} \begin{cases} \alpha \text{ is surjective} \\ \beta \text{ is surjective} \end{cases} \Rightarrow$$

$$\begin{cases} \forall f_i \in \mathcal{A} \Rightarrow\ f_i = f(i), \exists\, i \in \mathbb{N} \\ \forall g_i \in \mathcal{Z} \Rightarrow\ g(i) = g_i, \exists\, e \in \mathbb{N} \end{cases}. \tag{2.2}$$

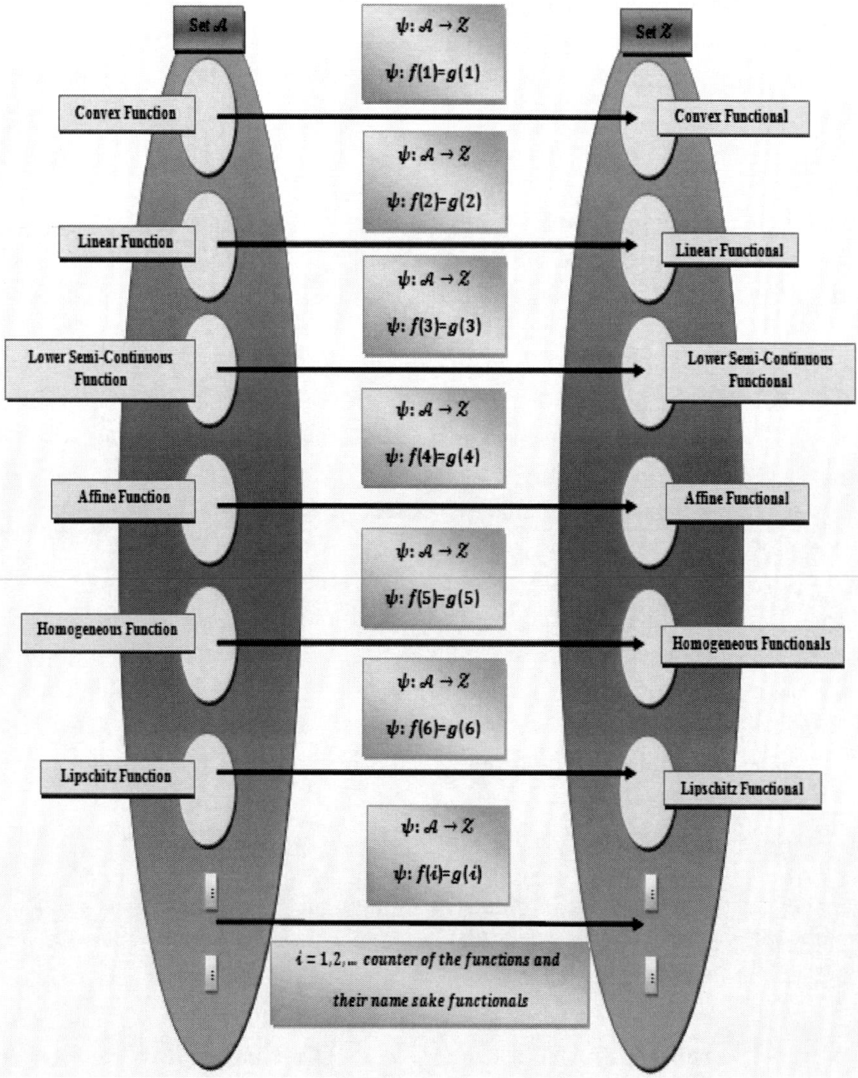

Figure 2.1. The existence of bijective maps from the set of all functions defined on $\mathbb{R}^n$ to the set of all functionals defined on $\mathbb{R}^n$ via SRM.

Based on (2.1), (2.2) and Definition 1.3:

$$\begin{cases} \psi: \mathcal{A} \longrightarrow \mathcal{Z}: i \neq e \rightarrow f_i \neq g_e \rightarrow f(i) \neq g(e) \\ \psi: \mathcal{A} \longrightarrow \mathcal{Z}: \forall g_i \in \mathcal{Z} \Rightarrow \exists f_i \in \mathcal{A}: f_i = g_i \end{cases}$$

$$\begin{cases} \Rightarrow \psi \text{ is injective} \\ \Rightarrow \psi \text{ is surjective} \end{cases}$$

Then, ψ is bijective (see Figure 2.1)■.

Based on Definition 1.4, let Θ be a predicate that performs the existence of the bijective maps $\psi_{k}: \mathcal{A} \longrightarrow \mathcal{Z}$, which connect the similar elements of the sets together (where $k \in \mathbb{N}$). Based on the construction of the sets $\mathcal{A}$ and $\mathcal{Z}$, it is clear while the sets $\mathcal{A}$ and $\mathcal{Z}$ are empty then $\Theta(0)$ is true, because it is also empty. And while the sets $\mathcal{A}$ and $\mathcal{Z}$ have one element i.e., $\mathcal{B}_1$ (set of all convex functions defined on $\mathbb{R}^n$) and $\mathcal{B}_1^*$ (set of all convex functionals defined on $\mathbb{R}^n$) then $\Theta(1)$ is true because there is a bijective map ψ_1 which connects both of them together based on Definition 1.3 and Theorem 2.1. Also while the sets $\mathcal{A}$ and $\mathcal{Z}$ have two elements i.e., $\mathcal{B}_1$ (set of all convex functions defined on $\mathbb{R}^n$), $\mathcal{B}_2$ (set of all linear functions defined on $\mathbb{R}^n$) and $\mathcal{B}_1^*$ (set of all convex functionals defined on $\mathbb{R}^n$), $\mathcal{B}_2^*$ (set of all linear functionals defined on $\mathbb{R}^n$) then $\Theta(2)$ is true because there are two bijective maps ψ_1 and ψ_2 which connect $\mathcal{B}_1$ to $\mathcal{B}_1^*$ and $\mathcal{B}_2$ to $\mathcal{B}_2^*$ respectively based on Definition 1.3 and Theorem 2.1. It means that $\Theta(2)$ is true because $\Theta(1)$ is true. By the continuity of the process, it is also clear that $\Theta(3)$ is true because $\Theta(2)$ is true. Finally, $\Theta(k)$ is true because $\Theta(k-1)$ is true and also $\Theta(k+1)$ is true because $\Theta(k)$ is true. It induces that the conditions of Definitions 1.4 hold and all elements of the Set $\mathcal{A}$ is connected to their similar elements of the Set $\mathcal{Z}$ which have the same name i.e., $\mathcal{B}_{k}$ (affine functions on $\mathbb{R}^n$) and $\mathcal{B}_{k}^*$ (affine functionals on $\mathbb{R}^n$), (see Figure 2.1).

As a result of the above procedure the following remark will appear.

Remark 2.1(Generalizability Property): SRM demonstrates that if there is a type of function defined on $\mathbb{R}^n$; then there is its same-name functional on $\mathbb{R}^n$, i.e., if there is a linear function defined on $\mathbb{R}^n$then there is a linear functional on $\mathbb{R}^n$ (see Figure 2.1).

3. Inheritance of Function Property in Functionals

In this section, based on the following theorem the Inheritance of function property for functional on $\mathbb{R}^n$, is described. Moreover, it is clear that $\mathcal{Z}$ (the set of all functionals defined on $\mathbb{R}^n$) is a subset of $\mathcal{A}$ (the set of all functions defined on $\mathbb{R}^n$). In other words, based on the definition, when a functional is a type of function, then $\mathcal{Z} \subseteq \mathcal{A}$.

Theorem 3.1 (SUBSET PRINCIPLE): If $\Xi(x)$ is a condition on sets, then, for each set $\mathcal{A}$, there exists a unique set whose members are precisely those members $\mathcal{Z}$ of $\mathcal{A}$ for which $\Xi(\mathcal{Z})$ holds [6]. where $\Xi(.)$ is the Laws and Properties of functions.

Remark 3.1 (Inheritance Property): Based on Theorem 3.1, it is clear that subsets inherit their properties from their superset. It implies that any functional defined on $\mathbb{R}^n$, should observe all function laws and properties defined on $\mathbb{R}^n$ as its inherited property.

SRM, Theorem 2.1, Theorem 3.1 and bijectivity of the map ψ demonstrate that there exist the following results, such that they validate the existence of Inheritance and Generalizability properties from functions into functionals:

All functionals defined on $\mathbb{R}^n$, inherit their properties from their same-name functions on $\mathbb{R}^n$;

If there is a type of function defined on $\mathbb{R}^n$ then there is its same-name functional on $\mathbb{R}^n$, i.e., if there is a concave function defined on $\mathbb{R}^n$ then there is a concave functional on $\mathbb{R}^n$.

It is concluded from the above discussion that generally based on the Inheritance and Generalizibility properties the SRM proposes the following theorem.

Theorem 3.2: If and only if there is a definition for functions on $\mathbb{R}^n$ then there is its corresponding definition for functionals on $\mathbb{R}^n$.

Proof: Based on Procedure 2.1:

1. Describe two sets $\mathcal{U}_1$ as the set of all function definitions on $\mathbb{R}^n$ which is countable and indexed via SRM and $\mathcal{U}_2$ as the set of all functional definitions on $\mathbb{R}^n$ which is countable and indexed via SRM. A map ψ which was introduced from the set $\mathcal{U}_1$ into $\mathcal{U}_2$ is bijective based on SRM. It implies that the sets $\mathcal{U}_1$ and $\mathcal{U}_2$ are one-to-one correspondence and the set $\mathcal{U}_2$ inherits its properties from its superset $\mathcal{U}_1$ (see Figure 3.1).

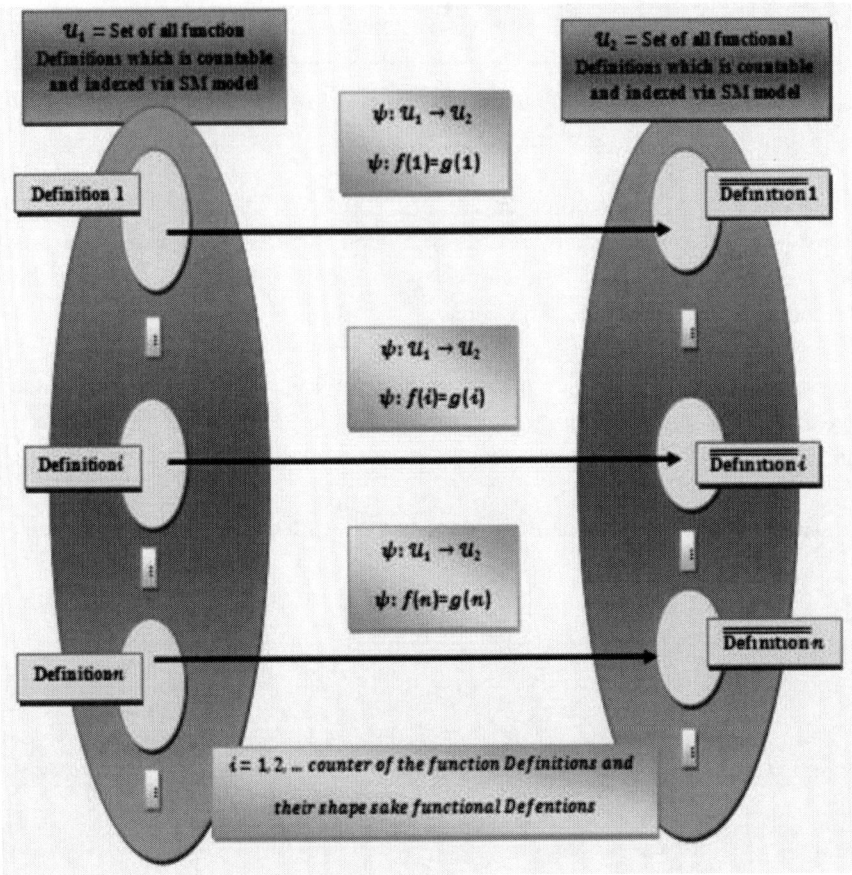

Figure 3.1. Existence of bijective maps from the set of all function definitions on $\mathbb{R}^n$ to the set of all functional definitions on $\mathbb{R}^n$ via SRM.

2. Also, let Θ be the statement that presents the existence of the bijective maps $\psi_{k}: \mathcal{U}_1 \longrightarrow \mathcal{U}_2$, connecting the similar elements of the sets together (where $k \in \mathbb{N}$). Based on construction of the sets $\mathcal{U}_1$ and $\mathcal{U}_2$, it is clear while the sets $\mathcal{U}_1$ and $\mathcal{U}_2$ have one element i.e., Δ_1 (Abstract convexity definition for functions on $\mathbb{R}^n$) and $\overline{\overline{\Delta_1}}$ (Abstract convexity definition for functionals on $\mathbb{R}^n$) respectively, then $\Theta(1)$ is true because there is a bijective map ψ_1 which connects both of them together based on Theorem 2.1. Also while the sets $\mathcal{U}_1$ and $\mathcal{U}_2$ have two elements i.e., Δ_1 (Abstract convexity definition for functions on $\mathbb{R}^n$), Δ_2 (Homogeneity definition for functions on $\mathbb{R}^n$) and $\overline{\overline{\Delta_1}}$

(Abstract convexity definition for functionals on $\mathbb{R}^n$), $\overline{\overline{\Delta_2}}$ (Homogeneity definition for functionals on $\mathbb{R}^n$) respectively, then $\Theta(2)$ is true because there are two bijective maps ψ_1 and ψ_2 which connect Δ_1 to $\overline{\overline{\Delta_1}}$ and Δ_2 to $\overline{\overline{\Delta_2}}$ respectively based on Theorem 2.1. It means that $\Theta(2)$ is true because $\Theta(1)$ is true. By induction, it is also clear that $\Theta(3)$ is true because $\Theta(2)$ is true, etc. Finally, $\Theta(\mathcal{k})$ is true because $\Theta(\mathcal{k}-1)$ is true and $\Theta(\mathcal{k}+1)$ is true because $\Theta(\mathcal{k})$ is true. It is understood that the condition of Definition 1.4 holds and all elements of the set $\mathcal{U}_1$ are connected to their similar elements of the set $\mathcal{U}_2$ which have the same name i.e., $\Delta_{\mathcal{k}}$ (abstract convexity definition for functions on $\mathbb{R}^n$) and $\overline{\overline{\Delta_{\mathcal{k}}}}$ (abstract convexity definition for functionals on $\mathbb{R}^n$).

Inheritance and Generalizability properties and Theorem 3.2 asserted that all the definitions of functions whih are defined on $\mathbb{R}^n$, can be expanded to propose new definitions for functionals on $\mathbb{R}^n$, i.e., abstract convexity definition for function on $\mathbb{R}^n$ can be extended to abstract convexity for functional on $\mathbb{R}^n$. The next section shall demonstrate the definition of abstract convex functional based on SRM in investigating the validity of the Theorem 3.2 experimentally.

4. OVERVIEW

First, some useful definitions and theorems shall be presented from functional analysis which will be required to be used in this section. According to these, abstract convex functionals and their properties shall be defined. The ultimate aim is to state the subdifferential of abstract convex functionals. Let $\overline{\mathbb{R}} = \mathbb{R} \cup \{-\infty\} \cup \{+\infty\}$. The following theorem is considred as a guide in this research to fulfill its chief aim which is definition of abstract convex functional.

Theorem 4.1 (*Hahn-Banach Theorem for Lower Semicontinuous Functionals*): *Let $\widetilde{\mathcal{X}}$ be normed vector space and suppose that the functional $F: \mathcal{X} \to \overline{\mathbb{R}}$, is proper, convex and lower semicontinuous. Then F is bounded below by an affine functional* [7].

The next definitions and theorems are extensions to propositions on functions which are stated in [8] as these propositions are required for the following discussion on the convex-along-rays functionals.

One of the chief findings of convex analysis exploited from the Theorem1 states that an arbitrary lower semicontinuous convex functional F (perhaps admitting the value $+\infty$) is the upper envelop of the set of all its affine minorants:

$$F(x) = \sup\{H(x) \text{is an affine functional}, F(x) \leq F(x) \text{ for all } x. \quad (4.1)$$

The supremum of the above set is obtained at a vector x if and only if the subdifferential of F at x is nonempty. Since affine functionals are defined by means of linear functionals, one can say that convexity is linearity + envelope representation. In particular, abstract convex with respect to $\mathcal{H}$ or $\mathcal{H}$-convex functionals can be represented as upper envelopes of subsets of a set $\mathcal{H}$ in abstract convexity theory (convexity without linearity). The set $\text{supp}(F, \mathcal{H}) = \{H \in \mathcal{H}, H \leq F\}$ of all $\mathcal{H}$-minorants of a functional F is called the support set (or $\mathcal{H}$-support set) of this functional. A set $\mathcal{U} \subset \mathcal{H}$ is called abstract convex or $(\mathcal{H}, \mathcal{X})$-convex if this set coincides with the support set of an abstract convex functional.

Two equal definitions of the subdifferentials for a convex functionals are available. The first definition is based on the global behavior of the functional. A linear functional L is called a subgradient (i.e., a member of the subdifferential) of the functional F at a vector y if the affine functional $H(x) = L(x) - (L(y) - F(y))$ is a support functional with respect to F, that is, $H(x) \leq F(x)$ for all x. On the other hand, the second definition has a local nature and related to a local approximation of the functional: the subdifferential is a closed convex set of linear functionals such that the directional derivative $F'_x(h) := \lim_{b \to +0} b^{-1}(F(x + bh) - F(x))$ is the upper envelope of this set. Both definitions demonstrate the support and tangent sides of the gradient for a differentiable convex functional, respectively.

5. Preliminaries

The functionals $F: \mathcal{X} \to \mathcal{V}$ shall be studied where $\mathcal{X}$ is a set and $\mathcal{V} \subset \overline{\mathbb{R}}$ is a segment. Segment $\mathcal{V}$ as a rule will conform to one of the following segments:

$$\mathbb{R} = (-\infty, +\infty), \overline{\mathbb{R}} = [-\infty, +\infty], \mathbb{R}_{+\infty} = (-\infty, +\infty] \quad (5.1)$$

$$\text{and } \mathbb{R}_+ = [0, +\infty), \overline{\mathbb{R}}_+ = [0, +\infty], \mathbb{R}_{++} = (0, +\infty), \overline{\mathbb{R}}_{++} = (0, +\infty]. \quad (5.2)$$

Let $\mathcal{V} \subset \overline{\mathbb{R}}$ be a segment. We shall investigate the supremum and the infimum of each subset of $\mathcal{V}$, specifically, supremum and the infimum of the empty subset $\emptyset$ of $\mathcal{V}$. It is appropriate to admit the following convention:

$$\inf \emptyset = \sup \mathcal{V}, \ \sup \emptyset = \inf \mathcal{V}. \tag{5.3}$$

Especially, $\inf \emptyset = +\infty$ for every $\mathcal{V}$ from (5.1) and (5.2). If $\mathcal{V}$ conform to one of the sets (5.1) then $\sup \emptyset = -\infty$; if $\mathcal{V}$ is one of the sets (5.2), then $\sup \emptyset = 0$. Let $\mathcal{V} \subset \overline{\mathbb{R}}$ be a segment and let $\mathcal{H}$ be a set of functionals $H: \mathcal{X} \to \mathcal{V}$. Assume that

$$\cup_{H \in \mathcal{H}} \{H(x) : x \in \mathcal{X}\} = \mathcal{V} \tag{5.4}$$

The pointwise supremum and the pointwise infimum of each subset of $\mathcal{H}$ shall be investigated. In particular, the pointwise supremum and infimum of the empty subset, $\emptyset$, of $\mathcal{H}$ will be investigated. The following convention is conforming to (5.3):

$$\sup_{H \in \emptyset} H(x) = \inf \mathcal{V}, \inf_{H \in \emptyset} H(x) = \sup \mathcal{V}, x \in \mathcal{X} \tag{5.5}$$

Therefore, the infimum of the empty subset of $\mathcal{H}$ is the constant functional equal to $\sup \mathcal{V}$ and the supremum of the empty set of $\mathcal{H}$ is the constant functional equal to $\inf \mathcal{V}$. Let $\mathcal{V} \subset \mathbb{R}$ be a segment and let $\bar{\mathcal{V}}$ be a closure of $\mathcal{V}$, that is,

$$\bar{\mathcal{V}} = \mathcal{V} \cup \{sup\ \mathcal{V}\} \cup \{inf\ \mathcal{V}\}. \tag{5.6}$$

Investigate a set $\mathcal{H}$ of functional $H: \mathcal{X} \to \mathcal{V}$. Let $\mathcal{U} \subset \mathcal{H}$ and $F_{\mathcal{U}}(x) = sup_{H \in \mathcal{U}} H(x)$. If $\mathcal{U}$ is nonempty, then the functional $F_{\mathcal{U}}$ maps into $\mathcal{V} \cup \{sup\ \mathcal{V}\}$. If $\mathcal{U}$ is empty, then the functional $F(x) = inf\ \mathcal{V}$ for all $x \in \mathcal{X}$. Therefore, $F_{\mathcal{U}}: \mathcal{X} \to \bar{\mathcal{V}}$ for each $\mathcal{U} \subset \mathcal{H}$.

6. Abstract Convex Functionals and Sets

This section presents the main definitions related to abstract convexity as extensions to propositions in [8] on functions in order to describe the CAR functionals. First, the following useful definition is explained.

Definition 6.1: Let $\mathcal{X}$ be a set, $\mathcal{V} \subset \bar{\mathbb{R}}$ and let $\mathcal{H}$ be a nonempty set of functionals $H: \mathcal{X} \to \mathcal{V}$. Respecting H (or H-convex) a functional $F: \mathcal{X} \to \bar{\mathcal{V}}$ is called abstract convex if a set $\mathcal{U} \subset \mathcal{H}$ such that F is the upper envelop of this set:

$$F(x) = \sup\{H(x): \mathrm{H} \in \mathcal{U}\}\, for\ all\ x \in \mathcal{X} \tag{6.1}$$

exists. Let $\mathcal{L}$ be the set of all linear functionals defined on $\mathbb{R}^n$ and $\mathcal{H}$ be the set of all affine functionals defined on $\mathbb{R}^n$. The following definition is based on affine operators ([9], [10]) and the definition of functionals is specifically extended.

Definition 6.2: Let $\breve{\mathcal{X}}$ be a vector space and subspace of $\mathbb{R}^n$. A functional $F: \breve{\mathcal{X}} \to \mathbb{R}$ is affine if there is a linear functional $L: \breve{\mathcal{X}}^{\#} \to \mathbb{R}$ and q an arbitrary element of $\breve{\mathcal{X}}$ such that

$$F(x) = L(x) + q \tag{6.2}$$

For all x in $\breve{\mathcal{X}}$. Where $\breve{\mathcal{X}}^{\#}$ is a uniquely determined vector subspace of $\breve{\mathcal{X}}$. On the other hand, an affine functional is just a linear functional plus a translation. However, at this level some theorems need to be proved. In that case first of all, the following definitions and theorems are presented. Both the following theorems proved in [11] are useful to continue this discussion:

Theorem 6.1: *Let $\breve{\mathcal{X}}$ be a real linear topological space and subspace of $\mathbb{R}^n$, x_0 a vector of $\breve{\mathcal{X}}$ and $L(x)$ a continuous semi-norm on $\breve{\mathcal{X}}$. Then there exists a continuous linear functional F on $\breve{\mathcal{X}}$ such that $F(x_0) = L(x_0)$ and $|F(x)| \leq L(x)$ on $\breve{\mathcal{X}}$* [11].

Theorem 6.2: *Let $\breve{\mathcal{X}}$ be a real linear topological space and subspace of $\mathbb{R}^n$, x_0 a vector of $\breve{\mathcal{X}}$ and $L(x)$ a real continuous functional on $\breve{\mathcal{X}}$ such that $L(x + y) \leq L(x) + L(y)$ and $L(bx) = bL(x)$ for $b \geq 0$. Then there exists a continuous linear functional F on $\breve{\mathcal{X}}$ such that $F(x_0) = L(x_0)$ and $-L(-x) \leq F(x) \leq L(x)$ on $\breve{\mathcal{X}}$* [11].

The next theorem which is based on these theorems is introduced:

Theorem 6.3: *$\breve{\mathcal{X}}$ be a real linear topological space and subspace of $\mathbb{R}^n$, x_0 a vector of $\breve{\mathcal{X}}$ and $L(x)$ a real linear continuous functional on $\breve{\mathcal{X}}$ such that $L(x + y) \leq L(x) + L(y)$ and $L(\mathscr{b}x) = \mathscr{b}L(x)$ for $\mathscr{b} \geq 0$. Then there exists a continuous linear affine functional F on $\breve{\mathcal{X}}$ such that $F(x_0) = L(x_0)$ and $F(x) = L(x) - c$ on $\breve{\mathcal{X}}$ where $c \in \mathbb{R}$.*

Proof from Theorem 6.3 states that $-L(-x) \leq F(x) \leq L(x)$. $F(x) \leq L(x)$, implying that $F(x) = L(x) + q$,

where $q \in \breve{\mathcal{X}}$. If we use $q = -c$ and $c \in \breve{\mathcal{X}}$. Thus,

$$F(x) = L(x) - c \tag{6.3}$$

From the Definition 6.2 and Equation (6.3) it is implied that for real linear continuous functional L, there exists a continuous linear affine functional F on $\breve{\mathcal{X}}$ in the form

$$F(x) = L(x) - c \;\blacksquare.$$

Now based on previous theorems, the next theorem is introduced as an important tool to define abstract convex functionals.

Theorem 6.4: *Let $\mathcal{L}$ be the set of all linear functionals defined on $\mathbb{R}^n$ and $\mathcal{H}$ be the set of all affine functionals defined on $\mathbb{R}^n$. A functional H is affine if and only if $H(x) = L(x) - c$, where $L \in \mathcal{L}$ and $c \in \mathbb{R}$.*

Proof: While H is affine functional then From Definition 2, Equation (9) and $q = -c$ $(c \in \mathbb{R})$: $H(x) = L(x) + q \Rightarrow H(x) = L(x) - c$. Conversely, while $H(x) = L(x) - c$. Let $q = -c$ $(c \in \mathbb{R})$ then $H(x) = L(x) + q$. Based on Definition 2 H is affine functional $\blacksquare$.

The next theorem is stated.

Theorem 6.5: *A function $F\colon \mathbb{R}^n \to \mathbb{R}_{+\infty}$ is $\mathcal{H}$-convex if and only if F is a lower semicontinuous convex function.*

Proof: Equation (4.1) implies that left to right condition of the theorem is correct and Theorem 4.1 infers that right to left condition of the theorem is correct. On the other hand, it is directly inferred from Definition 4.1 that if and only if $F(x) = \sup\{H(x): H \in \mathcal{H}, H \leq F, x \in \mathcal{X}\}$, where $H \leq F \Leftrightarrow (H(x) \leq F(x)$ for all $x \in \mathcal{X})$ ■,

F is $\mathcal{H}$-convex.

Remark 6.1: Then a functional $P: \mathbb{R}^n \to \mathbb{R}_{+\infty}$ is $\mathcal{L}$-convex if and only if P is lower semicontinuous sublinear functional.

Then, the following definition is defined.

Definition 6.3: Let $F: \mathcal{X} \to \bar{\mathcal{V}}$. The set

$$(F, H) = \{H \in \mathcal{H}, H \leq F\} \tag{6.4}$$

of all $\mathcal{H}$-minorants of F is called the support set of the functional F with respect to the set of elementary functional H. Sometimes, it is better to use term lower support set instead of support set.

If L is the set of all linear functionals and P is a lower semicontinuous sublinear functional such that $P(0) < +\infty$. Then $P(0) = 0$ and

$$\text{supp}\,(P, L) = \{L \in \mathcal{L} : (\forall x \in \mathbb{R}^n) L(x) \leq P(x)\}$$
$$= \{L \in \mathcal{L} : (\forall x \in \mathbb{R}^n)\, L(x) - L(0) \leq P(x) - P(0)\} = \partial P(0),$$

where $P(0)$ is the subdifferential of the convex functional P at the origin. Note the following definition which is useful for proof of following lemma:

Definition 6.4: A set $\mathcal{U} \subset \mathcal{H}$ is called abstract convex with respect to $\mathcal{X}$ (or $(\mathcal{H}, \mathcal{X})$-convex) if there exists a functional $F: \mathcal{X} \to \overline{\mathbb{R}}$ such that $\mathcal{U} = supp(F, \mathcal{H})$.

The following lemma which is useful for considering the subdifferentiability of AC functionals is stated in next part.

Lemma 6.1: (Separation property) Let $\mathcal{H}$ be a set of elementary functionals defined on a set $\mathcal{X}$. Let $\mathcal{U}$ be a proper subset of $\mathcal{H}$, that is $\mathcal{U} \neq \emptyset$,

$\mathcal{U} \neq \mathcal{H}$. Then the set $\mathcal{U}$ is $(\mathcal{H}, \mathcal{X})$-convex if and only if for each $H \notin \mathcal{U}$ there exists $x \in \mathcal{X}$ such that

$$H(x) > sup_{H' \in \mathcal{U}} H'(x) \tag{6.5}$$

Proof: similar to proof of lemma for functions in [8], let $\mathcal{U}$ be a proper $(\mathcal{H}, \mathcal{X})$-convex set. Let $P_{\mathcal{U}}$ be the upper envelope of the set $\mathcal{U}$ and

$$P_{\mathcal{U}}(x) = \sup\{H'(x): H' \in \mathcal{U}\}.$$

It is inferred from the definition of abstract convex sets that $\mathcal{U} = \{H \in \mathcal{H} : H(x) \leq P_{\mathcal{U}}(x)\ for\ all\ x \in \mathcal{X}\}$. Thus if $H \notin \mathcal{U}$ there exists $x \in \mathcal{X}$ such that $H(x) > P_{\mathcal{U}}(x)$. Now assume that for each $H \notin \mathcal{U}$ one can find $x \in \mathcal{X}$ such that (6.5) holds. Consider the set $\text{supp}(P_{\mathcal{U}}, \mathcal{H})$. Clearly $\mathcal{U} \subset \text{supp}(P_{\mathcal{U}}, \mathcal{H})$. If there exists $H \in \text{supp}(P_{\mathcal{U}}, \mathcal{H})$ such that $H \notin \mathcal{U}$ then applying (6.5) and x' can be fined such that $H(x') > P_{\mathcal{U}}(x')$, which is impossible since $H(x) \leq P_{\mathcal{U}}(x)$ for all $x \in \mathcal{X}$ ■.

7. SUBDIFFERENTIABILITY

As it was mentioned before, local and global are two different approaches to subdifferentiability of convex functionals. The global definition can be naturally extended to abstract convex setting. This section involves introduction and extensions to appropriate definitions and propositions for functionals which the reference [8] has for functions. Note that this section shall investigate only simple (finite elementary) functionals. Now the definition of the abstract subdifferential for abstract convex functionals is given.

Definition 7.1: Let $\mathcal{L}$ be a set of simple functionals defined on a set $\mathcal{X}$. A functional $L \in \mathcal{L}$ is called an abstract subgradient (or $\mathcal{L}$-subgradient) of a proper functional $F: \mathcal{X} \to \mathbb{R}_{+\infty}$ at a vector y if $F(x) \geq L(x) - (L(y) - F(y))$ for all $x \in \mathcal{X}$. The set $\partial_{\mathcal{L}} F(y)$ of all abstract subgradient of F at y is referred to as the abstract subdifferential (or $\mathcal{L}$-subdifferential) of the functional F at the vector y.

While the subdifferential $\partial_{\mathcal{L}}F(y)$ is not empty, it implies that $y \in dom\, F = \{x \in \mathcal{X}: F(x) < +\infty\}$. This assertion follows directly from Definition 7.1. Let $\mathcal{H}_{\mathcal{L}}$ be the closure of the set $\mathcal{L}$ under vertical shifts, that is

$$\mathcal{H}_{\mathcal{L}} = \{H': H'(x) = H(x) - c, H \in \mathcal{H}\}, c \in \mathbb{R}\}.$$

Clearly, $L \in \partial_{\mathcal{L}}F(y)$ if and only if $H_L \in supp(F, \mathcal{H}_{\mathcal{L}})$, where the functional

$$H_L(x) = L(x) - (L(y) - F(y)) \tag{7.1}$$

is an element of $\mathcal{H}_{\mathcal{L}}$. The following assertion holds:

Proposition 7.1: Let $F: \mathcal{X} \to \mathbb{R}_{+\infty}$ be a proper functional. Then the subdifferential $\partial_{\mathcal{L}}F(y)$ is nonempty if and only if

$$F(y) = \max\{H(y): H \leq F, H \in \mathcal{H}_{\mathcal{L}}\}. \tag{7.2}$$

Proof: Similar to proof for function in [8], let $L \in \partial_{\mathcal{L}}F(y)$ and let H_L be a functional defined by (7.1). Then $H_L \in \mathcal{H}_{\mathcal{L}}$, $H_L \leq F$ and $H_L(y) = F(y)$. Hence (14) holds. Now assume that there exists functional $H \in supp(F, \mathcal{H}_{\mathcal{L}})$ such that $F(y) = H(y)$. Let $H(x) = L(x) - c$ with $L \in \mathcal{L}$ and $c \in \mathbb{R}$. Then $c = L(y) - F(y)$, hence the inequality $H(x) - (L(y) - F(y)) \leq F(x)$ holds ■.

Corollary 7.1: If a set $\mathcal{L}$ is closed under vertical shifts, that is $\mathcal{L} = \mathcal{H}_{\mathcal{L}}$ then

$$\partial_{\mathcal{L}}F(y) = \{L \in supp(F, \mathcal{L}): L(y) = F(y)\}. \tag{7.3}$$

Let $\mathcal{L}$ be the set of linear functionals defined on a vector space $\mathcal{X}$ then the set $\mathcal{H}_{\mathcal{L}}$ consists of affine functionals. Considering this observation, the initial set $\mathcal{L}$ as a set of abstract linear functionals shall be investigated. Vertical shifts of functionals $L \in \mathcal{L}$, that is, functionals of the form

$$H(x) = L(x) - c \text{ for all } x \in \mathcal{X} \tag{7.4}$$

With $L \in \mathcal{L}$ and $c \in \mathbb{R}$ are called $\mathcal{L}$-affine functionals or abstract affine functionals with respect to $\mathcal{L}$. Therefore, the set $\mathcal{H}_{\mathcal{L}}$ is referred to as the set of $\mathcal{L}$-affine functionals. By property of (7.4) H and the pair $(L, c) \in \mathcal{L} \times \mathbb{R}$ is identified. It is possible that there are different pairs (L_1, c_1) and (L_2, c_2) such that $H(x) = L_1(x) - c_1 = L_2(x) - c_2$. For instance, if $\mathcal{L}$ is already closed under vertical shifts and $H(x) = L(x) - c$ then also is $H(x) = L'(x) - c'$ with $L' = L + d \in \mathcal{L}$, $c' = c + d \in \mathbb{R}$, where $1(x) = 1$ for all $x \in \mathcal{X}$ and $d \in \mathbb{R}$. Assume that the following condition (Σ) holds:

$$(\Sigma)\ L - c \notin \mathcal{L} \text{ for all } L \in \mathcal{L} \text{ and } c \neq 0 \tag{7.5}$$

It is inferred directly from (Σ) and it is possible to represent an $\mathcal{L}$-affine functional H in the form (7.4) in a unique way. In such a case the set $\mathcal{H}_{\mathcal{L}}$ of all -affine functionals can be identified with $\mathcal{L} \times \mathbb{R}$. Let $\mathcal{L}$ be a set of abstract linear functionals defined on $\mathcal{X}$ and let $F: \mathcal{X} \to \mathbb{R}_{+\infty}$ be a proper functional. Assume that the subdifferential $\partial_{\mathcal{L}} F(x) \neq \emptyset$ for all $x \in \mathcal{X}$. Then, Proposition 7.1 induces that the functional F is abstract convex with respect to the set $\mathcal{H}_{\mathcal{L}}$ of abstract affine functionals and the supremum $\sup\{H(x): H \in supp(F, \mathcal{H}_{\mathcal{L}})\}$ is attained for each $x \in \mathcal{X}$. This observation establishes that the $\mathcal{L}$-subdifferential is an appropriate tool in the study of $\mathcal{H}_{\mathcal{L}}$-convex functionals.

Definition 7.2: A subset $\mathcal{X}$ of a vector space is called a cone set if $\mathscr{b}x \in \mathcal{X}$ for each $x \in \mathcal{X}$ and $\mathscr{b} > 0$.

The following proposition informs a very simple explanation of the $\mathcal{L}$-subdifferential for a set $\mathcal{L}$ consisting of positively homogeneous functionals.

Proposition 7.2: Let $\mathcal{L}$ be the set of abstract linear functionals defined on a conic set $\mathcal{X}$ and let each $L \in \mathcal{L}$ be positively homogeneous of degree one $(L(\mathscr{b}x) = \mathscr{b}L(x)\ for\ all\ \mathscr{b} > 0)$. Then for an $\mathcal{L}$-convex functional P and for $\mathscr{y} \in \mathcal{X}$ we have:

$$\partial_{\mathcal{L}} P(\mathscr{y}) = \{L \in supp(P, \mathcal{L}): L(\mathscr{y}) = P(\mathscr{y})\} \tag{7.6}$$

Proof: Similar to proof on functions in [8], let $L \in \partial_{\mathcal{L}} P(\mathscr{y})$, that is,
$P(x) \geq L(x) - (L(\mathscr{y}) - P(\mathscr{y}))$ for every $x \in \mathcal{X}$. Let $x \in \mathcal{X}$. Investigate vectors $\mathscr{b}x$ with $\mathscr{b} > 0$. Then

$$\mathscr{b}P(x) = P(\mathscr{b}x) \geq \mathscr{b}L(x) - (L(y) - P(y)) \tag{7.7}$$

so,

$$\mathscr{b}P(x) = L(x) - \frac{L(y)-P(y)}{\mathscr{b}}.$$

Passing to the limit as $\mathscr{b} \to +\infty$, then $P(x) \geq L(x)$ is gotten, that is, $L \in supp(P, \mathcal{L})$. Specially, $L(y) \leq P(y)$ is stated. On the other hand, passing to the limit in (7.7) as $\mathscr{b} \to +\infty$, $L(y) \geq P(y)$ is gotten. Therefore, $L(y) = P(y)$ and $\partial_{\mathcal{L}} P(y)$ is contained in the set on the right-hand side in (7.6). Assume now that $P(x) \geq L(x)$ for all $x \in \mathcal{X}$ and $P(y) - L(y) = 0$. Then

$$P(x) \geq L(x) - (L(y) - P(y)) \text{ for all } x \in \mathcal{X} \blacksquare.$$

Remark 7.1: A similar result is valid if $\mathcal{L}$ is closed under vertical shifts (see Corollary 7.1).

CONCLUSION

The main finding of this paper supports the existence of Inheritance and Generalizability of function properties on functionals. Additionally, it was revealed that the Generalizability property encourage this fact that if there is any type of function defined on $\mathbb{R}^n$ then there is its same-name functional on $\mathbb{R}^n$. What is more, Inheritance property validates that the properties of function are inherited in its same-name functional and SRM is a major tool by which these properties are illustrated. This paper also introduced that the extension of SRM based on indexed family and Axiom of Induction demonstrates the existence of the properties in definitions of both functionals and functions. Consequently, all the definitions of functions can be extended to introduce the same new definitions on functionals. Finally, this study proposes the CAR functionals and their subdifferentiability based on these properties on functionals.

REFERENCES

[1] E. W. Weisstein, Functional Analysis, MathWorld--A Wolfram Web Resource, (1999). http://mathworld.wolfram.com/Functional Analysis. html.

[2] R.. Todd, Functional, From MathWorld--*A Wolfram Web Resource*, created by Eric W. Weisstein, (1999), http://mathworld. wolfram.com/ Functional.html.

[3] S. Lang, *Algebra*, Addison-Wesley, (1993).

[4] J. R. Munkres, *Topology*, United States of America: Prentice Hall, Inc. upper saddle River, (2000).

[5] V. J. Katz, *History of Mathematics: An Introduction*, Addison-Wesley, (1998).

[6] M.O. Searcoid, *Elements of Abstract Analysis*, Springer undergraduate series, Berlin Heidelberg New York: Springer-Verlag, (1948).

[7] A. J. Kurdila, and M. Abarankin, *Convex Functional Analysis*, Sestems & Control: Foundations and Applications, Birkhäuser Basel, (2005).

[8] A.M. Rubinov, *Abstract Convexity and Global Optimization*. Nonconvex Optimization and its Applications, Kluwer Academic Publishers, (2000).

[9] S. Boid, L. E. Ghaoui, E. Feron, and V. Balakrishnan, *Linear Matrix Inequalities in System and Control Theory*, SIAM Studies in Applied Mathematics, Philadelphia: Society for Industrial and Applied Mathematics, (1994).

[10] J. J. Koliha, and A. P. Leumg, *Ergodic Families of Affine Operators*, Springer-Verlage, Math.Ann. 216, (1975).

[11] K. Yosida, *Functional Analysis*, Berlin Heidelberg New York: Springer-Verlag, (1980).

In: A Note on Inheritance …
Editor: Iraj Sadegh Amiri
ISBN: 978-1-63485-784-0

Chapter 2

CONVEX-ALONG-RAYS FUNCTIONALS DEFINED ON $\mathbb{R}^n$ AND THEIR SUBDIFFERENTIABILITY

***Seyedalireza Seyedi*[1,*], *Iraj Sadegh Amiri*[2,†] *and Sara Chaghervand*[3,‡]**

[1]Department of Mathematics, Hamedan Branch,
Islamic Azad University, Hamedan, Iran
[2]Photonics Research Centre, University of Malaya,
Kuala Lumpur, Malaysia
[3]Department of Electrical Engineering, Borujerd Branch,
Islamic Azad University, Borujerd, Iran

ABSTRACT

As a major branch of Euclidean geometry, convexity has its proof in the field of algebra and analysis. This study investigates convex-along-rays functionals and based on the Inheritance and Generalizability of function properties on functionals it analyses the existence of their subdifferentiability which is defined on $\mathbb{R}^n$.

---

* Seyedalireza Seyedi: E-mail: seyedalireza.seyedi@gmail.com.
† Corresponding author: Iraj Sadegh Amiri. E-mail: amiri@um.edu.my.
‡ Sara Chaghervand: E-mail: engsara1@yahoo.com.

Keywords: abstract convexity, convexity-along-ray, subdifferential, functional

1. INTRODUCTION

Functional analysis, as a branch of mathematics, deals with vector spaces which have infinite dimensions as well as mappings existing between them. These mappings are called functionals on condition that the range is on the real line [1]. To analyse the properties of functionals based on the framework of functions, first some fundamental information about functions shall be presented. This information is collected from abstract analysis and topology and is presented as some definitions and theorems which are considered essential to be used in the current paper. Based on these, the study shall investigate the Inheritance and Generalizability of function properties for functionals. Then the study proposes Increasing Positively Homogeneous (IPH) functional on $\mathbb{R}^n$ based on function framework and those function properties.

A functional is traditionally a map which usually ranges from a vector space of functions to real numbers. To clarify this, usually a function takes functions as an argument or input and then returns a real number [2, 3]. The following examples show how Theorem 2.1 is proved through this paper.

Definition 1.1: A set is said to be finite if there is a bijective correspondence of $\mathcal{A}$ with some segments of the positive integers. That is, $\mathcal{A}$ is finite if it is empty or if there is a bijection $f: \mathcal{A} \to \{1, \dots, n\}$ [4].

Definition 1.2: A set $\mathcal{A}$ is said to be countably infinite if there is a bijective correspondence $f: \mathcal{A} \to \mathbb{Z}_+$ [4].

Let $\mathcal{A}$ be a set of all functions defined on $\mathbb{R}^n$. A map is described from the set of natural numbers to set $\mathcal{A}$ which implies that set $\mathcal{A}$ is countable. In the present discussion, the following theorems and corollary are regarded as really vital.

Theorem 1.1: Let $\mathcal{A}$ be a nonempty set. Then the following are equivalent [4]:

- A is countable;
- There is a surjective function $\boldsymbol{f}: \mathbb{Z}_+ \longrightarrow \boldsymbol{\mathcal{A}}$;
- There is an injective function $\boldsymbol{g}: \boldsymbol{\mathcal{A}} \longrightarrow \mathbb{Z}_+$.

Theorem 1.2: A countable union of countable sets is countable [4].

Corollary 1.1: A subset of a countable set is countable [4].

Definition 1.3 (Indexed Family): Let $\mathcal{F}$ be a nonempty collection of sets. An indexing function for $\mathcal{F}$ is a surjective function f from set $\mathcal{E}$, called the index set, to $\mathcal{F}$. The collection $\mathcal{F}$, together with the indexing function f, is called an indexed family of sets. Given $d \in \mathcal{E}$, the set $f(d)$ shall be denoted by the symbol f_d. And the indexed family itself shall be denoted by the symbol$\{ f_d\}_{d\in\mathcal{E}}$. Suppose that $f: \mathcal{E} \longrightarrow \mathcal{F}$ is an indexing function for $\mathcal{F}$; Let f_d denote $f(d)$, [4]. Then,

$$\bigcup_{d\in\mathcal{J}} f_d = \{x | \text{for at least one } d \in \mathcal{J}, x \in f_d\}.$$

Definition 1.4 (Axiom of Induction): The basic assumption of induction is, in logical symbols,

$$\forall \text{ predicates } \Theta, (\Theta(0) \wedge \forall k \in \mathbb{N}[\Theta(k) \Rightarrow \Theta(k+1)]) \Rightarrow \forall n \in \mathbb{N}\, \Theta(n)$$

where Θ is the proposition in question and k in addition to n are both natural numbers. In other words, the basis $\Theta(0)$ being true along with the inductive case ("$\Theta(k)$ is true implies $\Theta(k+1)$ is true for all natural k") being true together imply that $\Theta(n)$ is true for any natural number n [5].

2. Generalizability of Function Property for Functionals on $\mathbb{R}^n$

As it is understood from Definitions 1.3 and 1.4, in order to investigate the existence of the Inheritance and Generalizability properties, the procedure to construct a model shall be put foreword. This model conventionally is called the Seyedi-Rohanin model (SRM):

Procedure 2.1 (SRM):

Categorize set $\boldsymbol{\mathcal{A}}$ into all their subset groups via their names, i.e., let $\boldsymbol{\mathcal{B}_1}$ be a set of all convex functions on $\mathbb{R}^n$, $\boldsymbol{\mathcal{B}_2}$ be a set of all linear functions on $\mathbb{R}^n$, etc. Then index all these subset groups via indexed family (see Definition 1.3) i.e., $\boldsymbol{f_{d_1}}$ for the set $\boldsymbol{\mathcal{B}_1}$, $\boldsymbol{f_{d_2}}$ for the set $\boldsymbol{\mathcal{B}_2}$, etc. Then,

$$\alpha: f(d) = f_d,\ \alpha: \mathbb{Z}_+ \longrightarrow \mathcal{A}$$

Then,
$d \neq c \rightarrow f_d \neq f_c \rightarrow f(d) \neq f(c)$, implies α is injective and corange(α) $= \bigcup_1^\infty \mathcal{B}_m = \mathcal{A}$ implies that α are surjective, $\forall f_d \in \mathcal{A} \Rightarrow f_d = f(d), \exists\, d \in \mathbb{N}$. Thus, α is bijective.

The sets $\mathcal{B}_m$ are countably infinite ($m = 1,2,\ldots$). $\mathcal{A} = \bigcup_1^\infty \mathcal{B}_m$ is countably infinite, because of $\bigcup_1^\infty \mathcal{B}_m$ is the union of all countably infinite sets (see Theorem 1.2). It implies that $\mathbb{N}$ and set $\mathcal{A}$ have a one-to-one correspondence. Thus, it implies that set $\mathcal{A}$ is countably infinite (see Theorem 1.1).

Let $\boldsymbol{\mathcal{Z}}$ be a set of all functionals defined on $\mathbb{R}^n$. Similarly, set $\boldsymbol{\mathcal{Z}}$ is categorized into all their subsets groups via their names, i.e., convex functionals, linear functionals, etc. Then all these groups are indexed.

Describe a map from set $\boldsymbol{\mathcal{A}}$ to set $\boldsymbol{\mathcal{Z}}$ to demonstrate their one-to-one correspondence. The following theorem shall establish the bijectivity of the map which was described between the set $\boldsymbol{\mathcal{A}}$ and set $\boldsymbol{\mathcal{Z}}$ in the model.

Theorem 2.1: Let the set $\mathcal{A}$ be the set of all functions defined on $\mathbb{R}^n$ and set $\mathcal{Z}$ be the set of all functionals defined on $\mathbb{R}^n$. If (α, β) represents the bijective maps from the natural number to both sets, then ψ as a map from set $\mathcal{A}$ to $\mathcal{Z}$is bijective.

Proof: Assuming that α and β are bijective, it is implied that,

$$\begin{cases} \alpha \text{ is injective} \\ \beta \text{ is injective} \end{cases} \Rightarrow$$

$$\begin{cases} i \neq e \rightarrow f_i \neq f_e \rightarrow: \mathrm{f}(i) \neq: f(e) \\ i \neq e \rightarrow g_i \neq g_e \rightarrow: g(i) \neq: g(e) \end{cases} \tag{2.1}$$

$$\text{and} \begin{cases} \alpha \text{ is surjective} \\ \beta \text{ is surjective} \end{cases} \Rightarrow$$

$$\begin{cases} \forall f_i \in \mathcal{A} \Rightarrow f_i = f(i), \exists\, i \in \mathbb{N} \\ \forall g_i \in \mathcal{Z} \Rightarrow g(i) = g_i, \exists\, e \in \mathbb{N} \end{cases} \tag{2.2}$$

Based on (2.1), (2.2) and Definition 1.3:

$$\begin{cases} \psi: \mathcal{A} \longrightarrow \mathcal{Z}: i \neq e \rightarrow f_i \neq g_e \rightarrow f(i) \neq g(e) \\ \psi: \mathcal{A} \longrightarrow \mathcal{Z}: \forall g_i \in \mathcal{Z} \Rightarrow \exists\, f_i \in \mathcal{A}: f_i = g_i \end{cases}$$

$$\begin{cases} \Rightarrow \psi \text{ is injective} \\ \Rightarrow \psi \text{ is surjective} \end{cases}$$

Then, ψ is bijective (see Figure 2.1) ■.

Based on Definition 1.4, let Θ be a predicate performing the existence of the bijective maps $\psi_k: \mathcal{A} \longrightarrow \mathcal{Z}$, which connect the similar elements of the sets together (where $k \in \mathbb{N}$). Based on the construction of the sets $\mathcal{A}$ and $\mathcal{Z}$, it is clear while the sets $\mathcal{A}$ and $\mathcal{Z}$ are empty then $\Theta(0)$ is true, because it is also empty. And while the sets $\mathcal{A}$ and $\mathcal{Z}$ have one element i.e., $\mathcal{B}_1$ (set of all convex functions defined on $\mathbb{R}^n$) and $\mathcal{B}_1^*$ (set of all convex functionals defined on $\mathbb{R}^n$) then $\Theta(1)$ is true because there is a bijective map ψ_1 which connects both of them together based on Definition 1.3 and Theorem 2.1. Also, while the sets $\mathcal{A}$ and $\mathcal{Z}$ have two elements i.e., $\mathcal{B}_1$ (set of all convex functions defined on $\mathbb{R}^n$), $\mathcal{B}_2$ (set of all linear functions defined on $\mathbb{R}^n$) and $\mathcal{B}_1^*$ (set of all convex functionals defined on $\mathbb{R}^n$), $\mathcal{B}_2^*$ (set of all linear functionals defined on $\mathbb{R}^n$) then $\Theta(2)$ is true because there are two bijective maps ψ_1 and ψ_2 which connect $\mathcal{B}_1$ to $\mathcal{B}_1^*$ and $\mathcal{B}_2$ to $\mathcal{B}_2^*$ respectively based on Definition 1.3 and Theorem 2.1. It means that $\Theta(2)$ is true because $\Theta(1)$ is true. By the continuity of the process, it is also clear that $\Theta(3)$ is true because $\Theta(2)$ is true. Finally, $\Theta(k)$ is true because $\Theta(k-1)$ is true and also $\Theta(k+1)$ is true because $\Theta(k)$ is true. It induces that the conditions of Definitions 1.4 hold and all elements of the Set $\mathcal{A}$ are connected to their similar elements of the Set $\mathcal{Z}$ which have the same name i.e., $\mathcal{B}_k$ (affine functions on $\mathbb{R}^n$) and $\mathcal{B}_k^*$ (affine functionals on $\mathbb{R}^n$), (see Figure 2.1).

As a result of the above procedure, the following remark will appear.

Remark 2.1 (Generalizability Property): SRM indicates that if there is a type of function defined on $\mathbb{R}^n$; then there is its same-name functional on $\mathbb{R}^n$. I.e., if there is a linear function defined on $\mathbb{R}^n$then there is a linear functional on $\mathbb{R}^n$ (see Figure 2.1).

3. Inheritance of Function Property in Functionals

The Inheritance of function property for functional on $\mathbb{R}^n$ is illustrated based on the following theorem in this part. Furthermore, obviously $\mathcal{Z}$ (the set of all functionals defined on $\mathbb{R}^n$) is a subset of $\mathcal{A}$ (the set of all functions defined on $\mathbb{R}^n$. In other words, based on the definition when a functional is a type of function, then $\mathcal{Z} \subseteq \mathcal{A}$.

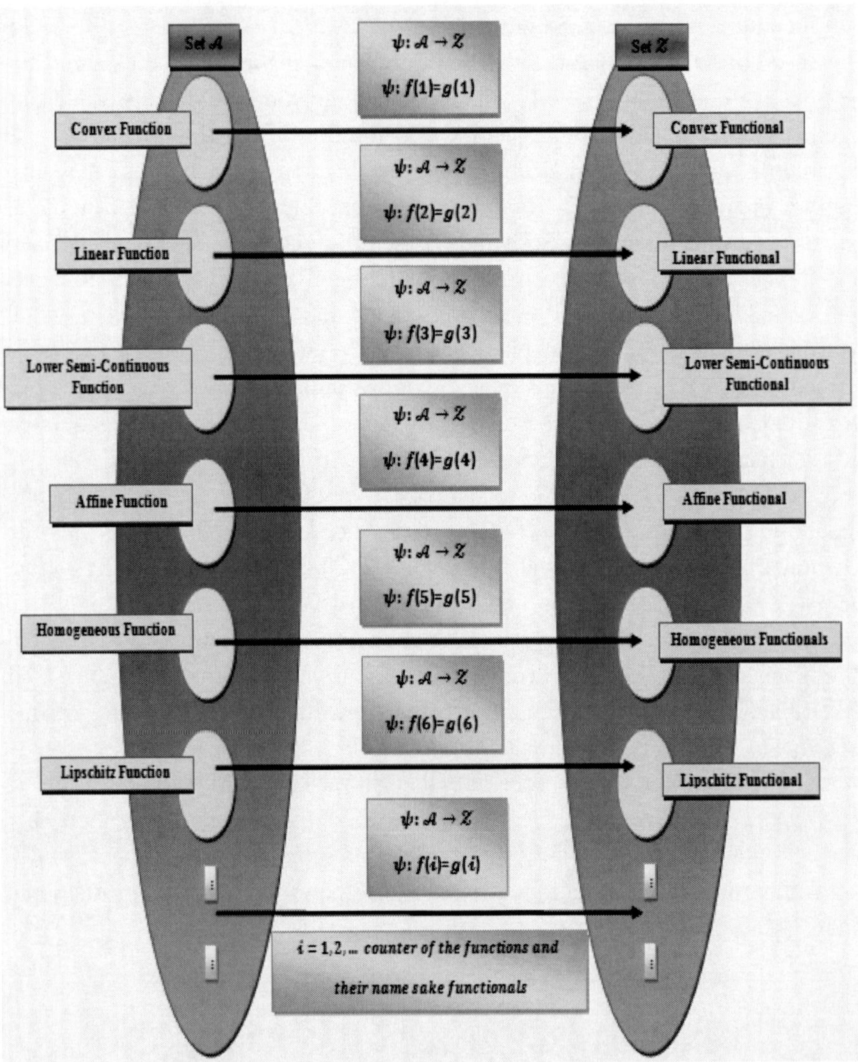

Figure 2.1. The existence of bijective maps from the set of all functions defined on $\mathbb{R}^n$ to the set of all functionals defined on $\mathbb{R}^n$ via SRM.

Theorem 3.1 (SUBSET PRINCIPLE): If $\Xi(x)$ is a condition on sets, then, for each set $\mathcal{A}$, there exists a unique set whose members are precisely those members $\mathcal{Z}$ of $\mathcal{A}$ for which $\Xi(\mathcal{Z})$ holds [6]. Where $\Xi(.)$ is the Laws and Properties of functions.

Remark 3.1 (Inheritance Property): Based on Theorem 3.1, it is clear that subsets inherit their properties from their superset. It implies that any functional defined on $\mathbb{R}^n$ should observe all function laws and properties defined on $\mathbb{R}^n$ as its inherited property.

SRM, Theorem 2.1, Theorem 3.1 and bijectivity of the map ψ indicate that the following findings exist; as a result they substantiate the existence of Inheritance and Generalizability properties from functions into functionals:

1) All functionals defined on $\mathbb{R}^n$ inherit their properties from their same-name functions on $\mathbb{R}^n$;
2) If there is a type of function defined on $\mathbb{R}^n$ then there is its same-name functional on $\mathbb{R}^n$, i.e., if there is a concave function defined on $\mathbb{R}^n$ then there is a concave functional on $\mathbb{R}^n$.

It is concluded from the above discussion that generally based on the Inheritance and Generalizibility properties the SRM proposes the following theorem.

Theorem 3.2: If and only if there is a definition for functions on $\mathbb{R}^n$ then there is its corresponding definition for functionals on $\mathbb{R}^n$.

Proof: Based on Procedure 2.1:

1) Describe two sets $\mathcal{U}_1$ as the set of all function definitions on $\mathbb{R}^n$ which is countable and indexed via SRM and $\mathcal{U}_2$ as the set of all functional definitions on $\mathbb{R}^n$ which is countable and indexed via SRM. A map ψ introduced from the set $\mathcal{U}_1$ into $\mathcal{U}_2$ is bijective based on SRM. It implies that the sets $\mathcal{U}_1$ and $\mathcal{U}_2$ are one-to-one correspondence and the set $\mathcal{U}_2$ inherits its properties from its superset $\mathcal{U}_1$ (see Figure 3.1).
2) Also, let Θ be the statement that present the existence of the bijective maps $\psi_{k}: \mathcal{U}_1 \longrightarrow \mathcal{U}_2$, which connect the similar elements of the sets together (where $k \in \mathbb{N}$). Based on construction of the sets $\mathcal{U}_1$ and $\mathcal{U}_2$, it is clear while the sets $\mathcal{U}_1$ and $\mathcal{U}_2$ have one element i.e., Δ_1 (Abstract convexity definition for functions on $\mathbb{R}^n$) and $\overline{\overline{\Delta_1}}$ (Abstract convexity definition for functionals on $\mathbb{R}^n$) respectively, then $\Theta(1)$ is true because there is a bijective map ψ_1 which connects both of them together based on Theorem 2.1. Also while the sets $\mathcal{U}_1$ and $\mathcal{U}_2$ have two elements i.e., Δ_1 (Abstract convexity definition for functions on $\mathbb{R}^n$), Δ_2 (Homogeneity definition for functions on $\mathbb{R}^n$) and $\overline{\overline{\Delta_1}}$ (Abstract convexity definition for functionals on $\mathbb{R}^n$), $\overline{\overline{\Delta_2}}$ (Homogeneity definition for functionals on $\mathbb{R}^n$) respectively, then

$\Theta(2)$ is true because there are two bijective maps ψ_1 and ψ_2 which connect Δ_1 to $\overline{\overline{\Delta_1}}$ and Δ_2 to $\overline{\overline{\Delta_2}}$ respectively based on Theorem 2.1. It means that $\Theta(2)$ is true because $\Theta(1)$ is true. Inductively, it is also clear that $\Theta(3)$ is true because $\Theta(2)$ is true, etc. Finally, $\Theta(\mathscr{k})$ is true since $\Theta(\mathscr{k}-1)$ is true and $\Theta(\mathscr{k}+1)$ is true because $\Theta(\mathscr{k})$ is true. It induces that the condition of Definition 1.4 holds and all elements of the set $\mathcal{U}_1$ are connected to their similar elements of the set $\mathcal{U}_2$ which have the same name i.e., $\Delta_{\mathscr{k}}$ (convexity-along-ray definition for functions on $\mathbb{R}^n$) and $\overline{\overline{\Delta_{\mathscr{k}}}}$ (convexity-along-ray definition for functionals on $\mathbb{R}^n$).

Inheritance and Generalizability properties and Theorem 3.2 asserted that new definitions for functionals on $\mathbb{R}^n$can be proposed through expanding all the definitions of functions defined on $\mathbb{R}^n$. I.e., convexity-along-ray definition for function on $\mathbb{R}^n$ can be broadened to convexity-along-ray for functional on $\mathbb{R}^n$. The next section shall indicate the definition of convex-along-ray functional based on SRM in investigating the validity of the Theorem 3.2 experimentally.

4. Convex-Along-Ray Functional Definition

In this section Convex-along-ray (CAR) functional is introduced based on the Inheritance and Generalizability properties of function on functional. Furthermore, this section proposes some theorems and propositions as stated in [7] and that they shall be broadened to functionals and the proofs of them are one-to-one correspondence to their proof in the form of function. At the first step the extension of convex-along-rays function definition in functionals is introduced.

Definition 4.1: Let $\mathcal{Q} \subset \mathbb{R}^n$ be a cone. A functional $F\colon \mathcal{Q} \to \mathbb{R}_{+\infty}$ is called convex-along-rays (CAR) if, for each $x \in \mathcal{Q}$, the functional

$$F_x(t) = F(tx)\; t \in [0,1) \tag{4.1}$$

is convex.

The closed ray $\mathcal{R}_x$ is denoted by the set: $\{tx\colon 0 \leq t < +\infty\}$. This definition is extended in the following definition.

Definition 4.2: A functional $F: \mathcal{Q} \to \mathbb{R}_{+\infty}$ is called convex-along-rays (CAR) if its restriction to each ray $\mathcal{R}_x$ with $x \in \mathcal{Q}$ is a convex functional.

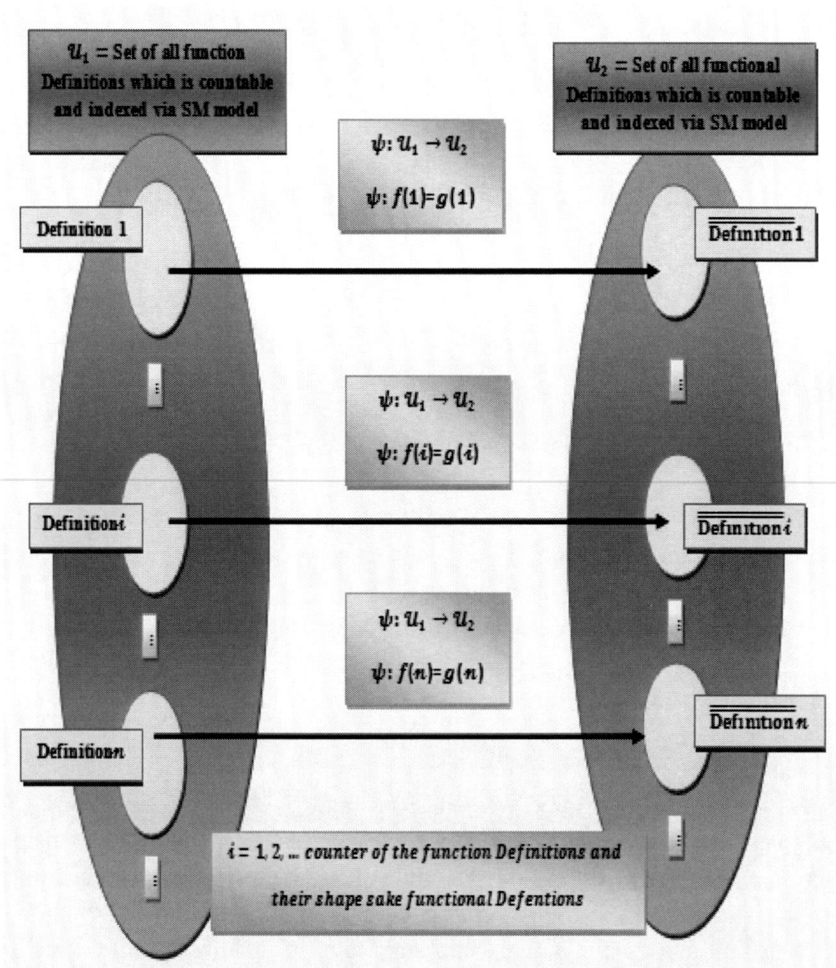

Figure 3.1. Existence of bijective maps from the set of all function definitions on $\mathbb{R}^n$ to the set of all functional definitions on $\mathbb{R}^n$ via SRM.

In this part, the increasing convex-along-rays (ICAR) functionals defined on $\mathcal{Q} = \mathbb{R}^n_+$ or $\mathcal{Q} = \mathbb{R}^n_{++}$ is investigated. Therefore, the following definition is stated.

Definition 4.3: Let $\mathcal{Q}$ be either $\mathbb{R}^n_+$ or $\mathbb{R}^n_{++}$. A functional $F: \mathcal{Q} \to \mathbb{R}_{+\infty}$ is an ICAR functional if the following conditions hold:

1) F is increasing: $x \geq y$ implies $F(x) \geq F(y)$;
2) for each $x \in Q$, the functional F_x defined by (4.1) is convex.

By continuing this way, some examples of ICAR functionals are:

a) An increasing convex functionals defined on $\mathbb{R}^n_+$ is ICAR.
b) An increasing positively homogeneous functional of degree $m \geq 1$ defined on $\mathbb{R}^n_+$ is ICAR. Especially,
   1. An IPH functional defined on $\mathbb{R}^n_+$ is ICAR.

The following propositions are found to be practical in order to show the properties of ICAR functionals.

Proposition 4.1: The set $\mathcal{F}$ of all ICAR functionals defined on either $\mathbb{R}^n_+$ or $\mathbb{R}^n_{++}$ is a convex cone.

Proof: Directly from the definition, it is easy to show that if F_1, F_2 is ICAR functionals and $b_1, b_2 > 0$ then the functional $b_1 F_1 + b_2 F_2$ is ICAR ■.

Proposition 4.2: Let F be an increasing functional on $\mathbb{R}^n_+$. Then the level sets $\{x \in \mathbb{R}^n_+ : F(x) \leq c\}$ are normal and the level sets $\{x \in \mathbb{R}^n_+ : F(x) \geq c\}$ are conormal. Specially, the set $dom\, F = \{x \in \mathbb{R}^n_+ : F(x) < +\infty\}$ is normal and the set $\{x \in \mathbb{R}^n_+ : F(x) = +\infty\}$ is conormal.

Proof: The proof follows directly from the definition 4.3 ■.

Proposition 4.3: Any $\mathcal{H}_\mathcal{L}$-convex functional $F: \mathbb{R}^n_+ \to \mathbb{R}_{+\infty}$ is lower semicontinuous and ICAR.

Proof: Let F be an $\mathcal{H}_\mathcal{L}$-convex functional, that is, $F(x) = \sup\,\{H(x): H \in supp\,(F, \mathcal{H}_\mathcal{L})\}$. Then for each $x \in \mathbb{R}^n_+$: $F_x(t) = \sup\{H_x(t): H \in supp(F, \mathcal{H}_\mathcal{L})\}$, where $F_x(t)$ and $H_x(t)$ are defined as in (1). Note that the functional H_x is affine for each $H \in \mathcal{H}_\mathcal{L}$ and $x \in \mathbb{R}^n_+$, hence the functional F_x is convex. Thus F is convex-along-rays. Since each functional $H \in \mathcal{H}_\mathcal{L}$ is increasing, it follows that F is increasing as well. Since each $H \in \mathcal{H}_\mathcal{L}$ is a continuous functional, implying that it is lower semicontinuous ■ which is the upper envelope of the set supp $(F, \mathcal{H}_\mathcal{L})$ of continuous functionals.

The next assertion gives this information that the class of ICAR functionals is enormous. But before that performance of one lemma is useful.

Lemma 4.1: Let $\mathcal{H}$ be a set of continuous functionals defined on a metric space $\breve{Z}$ with the following properties:

1. $\mathcal{H}$ is a conic set;

2. $H + b\mathbf{1} \in \mathcal{H}$ for each $H \in \mathcal{H}$ and $b < 0$ (here $1(x) = 1$ for all $x \in \breve{\mathcal{Z}}$);
3. For each $z \in \breve{\mathcal{Z}}$ there exists a functional $H \in \mathcal{H}$ such that $H(z) = 0$, $H(x) < 0$ for $x \neq z$, $H + \sigma\mathbf{1} \in \mathcal{H}$ for all small enough $\sigma > 0$.

Then a functional F defined on $\breve{\mathcal{Z}}$ is H-convex if and only if F is lower semicontinuous [7]. Thus, based on this lemma the following proposition shall be stated.

Proposition 4.4: Let F be a lower semicontinuous functional defined on the unit simplex $\mathcal{S}^* = \{x \in \mathbb{R}^n_+ : \sum_i x_i = 1\}$. Then there exists an ICAR extension of F, that is, an ICAR functional $\bar{F}: \mathbb{R}^n_+ \to \mathbb{R}_{+\infty}$ such that $\bar{F}(x) = F(x)$ for all $x \in \mathcal{S}^*$.

Proof: Similar to proof of functions let $\mathcal{H}_{\mathcal{S}^*}$ be the set of all functionals $H_{\mathcal{S}^*}$ defined on the simplex $\mathcal{S}^*$ by $H_{\mathcal{S}^*}(x) = \langle \ell, x \rangle - c$ with $\ell \in \mathbb{R}^n_+$, $c \in \mathbb{R}$ where $\langle \ell, x \rangle = \min_{i \in \mathcal{I}_+(l)} \ell_i x_i$ and $\mathcal{I}_+(\ell) = \{i : \ell_i > 0\}$. Obviously, conditions (i) and (ii) from Lemma 1 hold for the set $\mathcal{H}_{\mathcal{S}^*}$. In addition, the condition (iii) can be checked. Let $z \in \mathcal{S}^*$. Consider the vector $\ell = 1/z$, where

$$(\frac{1}{z})_i = \begin{cases} \frac{1}{z_i} \; if \; z_i > 0; \\ 0 \; if \; z_i = 0. \end{cases}$$

It is obvious that $\langle \ell, z \rangle = 1$. Since

$$\sum_{i=1}^{n} x_i = \sum_{i=1}^{n} z_i$$

for $x \in \mathcal{S}^*$, it induces that for $x \neq z$ there exists an index j such that $x_j < z_j$. Obviously, $j \in \mathcal{I}_+(z) = \mathcal{I}_+(\ell)$. Hence,

$$\langle \ell, x \rangle = \min_{i \in \mathcal{I}_+(z)} \frac{x_i}{z_i} < 1.$$

Consider the functional H defined on $\mathcal{S}^*$ by $H(x) = \langle \ell, x \rangle - 1$. Then

$H(z) = 0$ and $H(x) < 0$ for all $x \neq z$

Also,

$$H + \sigma.\mathbf{1} \in \mathcal{H}_{\mathcal{S}^*} \text{ for all } \sigma$$

Therefore, the condition (iii) from Lemma 4.1 holds. Let $F: \mathcal{S}^* \to \mathbb{R}_{+\infty}$ be a lower semicontinuous functional. It implies that from Lemma 4.1 there exists a set $\mathcal{U} \subset \mathcal{H}_{\mathcal{S}^*}$ such that $F(x) = sup_{H \in \mathcal{U}} H(x)$ for all $x \in \mathcal{S}^*$. Investigate now the functional $\tilde{F}$ defined on $\mathbb{R}_+^n$ by

$$\tilde{F}(x) = \sup\{H(x): H \in \mathcal{U}\}.$$

It follows from Proposition 4.3 that $\tilde{F}$ is an ICAR functional. Also,

$$\tilde{F}(x) = F(x) \text{ for all } x \in \mathcal{S}^* \blacksquare.$$

Then based on lower semicontinuous functionals, the following propositions stated the relationship between ICAR functional and $\mathcal{H}_{\mathcal{L}}$-convex functional.

Proposition 4.5: Let $F: \mathbb{R}_+^n \to \mathbb{R}_{+\infty}$ be an ICAR functional such that for $x \in \mathbb{R}_+^n$ the functional F_x defined by (4.1) is lower semicontinuous on $\mathbb{R}_+$. Then the functional F is abstract convex with respect to the set $\mathcal{H}_{\mathcal{L}}$ of all functionals H defined by

$$H(x) = \langle \ell, x \rangle - c \text{ and } \ell, x \in \mathbb{R}_+^n \tag{4.2}$$

Proof: Proof of the theorem in functionals is similar to functions in [7] completely ■.

The following theorem is investigated based on Propositions 4.3, 4.4 and 4.5.

Theorem 4.1: *Let $\mathcal{H}_{\mathcal{L}}$ be the class of all functionals H by (4.2). A functional $F: \mathbb{R}_+^n \to \mathbb{R}_{+\infty}$ is $\mathcal{H}_{\mathcal{L}}$-convex if and only if F is lower semicontinuous and ICAR.*

Proof: Let F be a lower semicontinuous ICAR functionals. Then the functional F_y is lower semicontinuous for each y. Thus Proposition 4.5 demonstrates that F is $\mathcal{H}_{\mathcal{L}}$-convex. The reverse assertation follows from Proposition 4.3 ■.

The following proposition will be useful for subdifferential of ICAR functionals.

Proposition 4.6: Let F ba an ICAR functional and $y \in \mathbb{R}^n_+ \setminus \{o\}$ be a vector, such that $(1+\varepsilon)y \in domF$ for some $\varepsilon > 0$. Then there exists $\ell \in \mathbb{R}^n_+$ with the property

$$\langle \ell, x \rangle - \langle \ell, y \rangle \leq F(x) - F(y) \text{ for all } x \in \mathbb{R}^n_+.$$

In particular, the vector $\ell = u/y$ with $u \in \partial F_y(1)$ enjoys this property.

Proof: Proof of the theorem in functionals is similar to functions in [7] completely ■.

5. Subdifferentiability of Increasing Convex-Along-Rays Functionals

With the respect to the set $\mathcal{H}_\mathcal{L}$ of all $\mathcal{L}$-affine functionals, abstract convexity is investigated, where $\mathcal{L}$ is the set of all min-type functionals defined on $\mathbb{R}^n_+$. According to Theorem 4.1 and Proposition 4.6, following proposition can be easily proved.

Proposition 5.1: Concerning $\mathcal{H}_\mathcal{L}$ if and only if F is lower semicontinuous and ICAR, a functional $F: \mathbb{R}^n_+ \to \mathbb{R}_{+\infty}$ will be an abstract convex.

Proof: It follows from Theorem 4.1 directly ■.

When this way continues, the $\mathcal{L}$-subdifferential $\partial F_\mathcal{L}(x)$ of an ICAR functionals F at a vector $x \in dom\ F$ shall be studied. Recall that

$$\partial F_\mathcal{L}(x) = \{\ell \in \mathcal{L}: \langle \ell, y \rangle - \langle \ell, x \rangle \leq F(y) - F(x)\}.$$

Then,

Theorem 5.1: Let F be an ICAR functional and $x \in \mathbb{R}^n_+ \setminus \{o\}$ such that $(1+\varepsilon)x \in domF$ for some $\varepsilon > 0$. Then the subdifferential $\partial F_\mathcal{L}(x)$ is not empty and

$$\{u/x : u \in \partial F_x(1)\} \subset \partial F_\mathcal{L}(x) \tag{5.1}$$

where $F_x(t) = F(tx)$.

Proof: the finding directly comes as a result of Proposition 5.1 ■.

Since F is increasing ,it follows that $o \in \partial F_{\mathcal{L}}(o)$ therefore, $\partial F_{\mathcal{L}}(o)$ is also nonempty. Let F be a finite ICAR functional. For $x \in \mathbb{R}_+^n$ the following vector is investigated

$$F^{\nabla}(x) = \frac{F'(x,x)}{x} \tag{5.2}$$

It is inferred immediately from Theorem 5.1 that $F^{\nabla}(x) \in \partial F_{\mathcal{L}}(x)$ for all $x \in \mathbb{R}_+^n$.

CONCLUSION

This paper revealed the existence of Inheritance and Generalizability of function properties on functionals. Furthermore, it was concluded this fact that if there is any type of function defined on $\mathbb{R}^n$ then there is its same-name functional on $\mathbb{R}^n$ is advocated by the Generalizability property. Additionally, Inheritance property validates that the properties of function are inherited in its same-name functional. These properties are proved by a major tool called SRM. It was also found that based on indexed family and Axiom of Induction the existence of the properties in definitions of both functionals and functions can be demonstrated by the extension of SRM. Consequently, the study indicated that all definitions of functions can be expanded to propose the new same definitions on functionals. What is more, in this the study based on these properties on functionals, the CAR functionals and their subdifferentiability were proposed.

REFERENCES

[1] E. W. Weisstein, Functional Analysis, MathWorld--A Wolfram Web Resource, (1999). http://mathworld.wolfram.com/FunctionalAnalysis.html.

[2] R.. Todd, Functional, From MathWorld--A Wolfram Web Resource, created by Eric W. Weisstein, (1999), http://mathworld.wolfram.com/Functional.html.

[3] S. Lang, *Algebra*, Addison-Wesley, (1993).

[4] J. R. Munkres, *Topology,* United States of America: Prentice Hall, Inc. upper saddle River, (2000).

[5] V. J. Katz, *History of Mathematics: An Introduction*, Addison-Wesley, (1998).

[6] M. O. Searcoid, Elements of Abstract Analysis, Springer undergraduate series, Berlin Heidelberg New York: Springer-Verlag, (1948).

[7] A. M. Rubinov, Abstract Convexity and Global Optimization. Nonconvex Optimization and its Applications, Kluwer Academic Publishers, (2000).

In: A Note on Inheritance ...
Editor: Iraj Sadegh Amiri

ISBN: 978-1-63485-784-0

Chapter 3

INCREASING POSITIVELY HOMOGENEOUS FUNCTIONAL DEFINED ON $\mathbb{R}^n$ AND THEIR SUBDIFFERENTIABILITY

***Seyedalireza Seyedi*[1,*], *Iraj Sadegh Amiri*[2,†] *and Sara Chaghervand*[3,‡]**

[1]Department of Mathematics, Hamedan Branch, Islamic Azad University,Hamedan, Iran
[2]Photonics Research Centre, University of Malaya, Kuala Lumpur, Malaysia
[3]Department of Electrical Engineering, Borujerd Branch, Islamic Azad University, Borujerd, Iran

ABSTRACT

The homogeneous functionals which are defined on the cone $\mathbb{R}^n$ are well presented and developed. This paper discusses increasing positively homogeneous (IPH) functionals and based on the Inheritance and Generalizability of function properties on functionals, it investigates the existence of their sub differentiability on $\mathbb{R}^n$.

---

* seyedalireza.seyedi@gmail.com
† Corresponding Author address: amiri@um.edu.my.
‡ engsara1@yahoo.com.

Keywords: increasing positive homogeneity, subdifferential, functional

1. INTRODUCTION

Functional analysis is a branch of mathematics dealing with vector spaces which have infinite dimensions and mappings that exist between them. These mappings are called functionals if and only if the range is on the real line. [1]. This paper investigates the properties of functionals based on the framework of functions. To go through the process, first some essential information about functions shall be presented from abstract analysis and topology by describing some definitions and theorems which this study shall need to use. Based on these, the paper shall analyze the Inheritance and Generalizability of function properties for functionals. In the next step the paper shall present Increasing Positively Homogeneous (IPH) functional on $\mathbb{R}^n$ based on function framework as well as these properties.

A functional is conventionally a map which usually ranges from a vector space of functions to real numbers. To more elaborate it, a function takes functions as an argument or input and then returns a real number. [2], [3]. In addition, this paper will demonstrate how Theorem 2.1 is corroborated through the following examples.

Definition 1.1: A set is said to be finite if there is a bijective correspondence of $\mathcal{A}$ with some segments of the positive integers. That is, $\mathcal{A}$ is finite if it is empty or if there is a bijection $f: \mathcal{A} \to \{1, \dots, n\}$ [4].

Definition 1.2: A set $\mathcal{A}$ is said to be countably infinite if there is a bijective correspondence $f: \mathcal{A} \to \mathbb{Z}_+$, [4].

Let $\mathcal{A}$ be a set of all functions defined on $\mathbb{R}^n$. A map is described from the set of natural numbers to set $\mathcal{A}$ which implies that set $\mathcal{A}$ is countable. The following theorems and corollary are crucial to the discussion.

Theorem 1.1: Let $\mathcal{A}$ be a nonempty set. Then the following are equivalent [4]:

- **A is countable;**
- **There is a surjective function $f: \mathbb{Z}_+ \to \mathcal{A}$;**

- **There is an injective function $g: \mathcal{A} \longrightarrow \mathbb{Z}_+$.**

Theorem 1.2: A countable union of countable sets is countable [4].

Corollary 1.1: A subset of a countable set is countable [4].

Definition 1.3 (Indexed Family): Let $\mathcal{F}$ be a nonempty collection of sets. An indexing function for $\mathcal{F}$ is a surjective function f from set $\mathcal{E}$, called the index set, to $\mathcal{F}$. The collection $\mathcal{F}$, together with the indexing function f, is called an indexed family of sets. Given $d \in \mathcal{E}$, the set $f(d)$ shall be denoted by the symbol f_d. And the indexed family itself shall be denoted by the symbol$\{ f_d\}_{d\in\mathcal{E}}$. Suppose that $f: \mathcal{E} \longrightarrow \mathcal{F}$ is an indexing function for $\mathcal{F}$; Let f_d denote $f(d)$, [4]. Then,

$$\bigcup_{d\in\mathcal{J}} f_d = \{x | \text{for at least one } d \in \mathcal{J}, x \in f_d\}.$$

Definition 1.4 (Axiom of Induction): The basic assumption of induction is, in logical symbols,

$$\forall \text{ predicates } \Theta, (\Theta(0) \wedge \forall k \in \mathbb{N}[\Theta(k) \Rightarrow \Theta(k+1)]) \Rightarrow \forall n \in \mathbb{N}\, \Theta(n)$$

where Θ is the proposition in question and k and n are both natural numbers. In other words, the basis $\Theta(0)$ being true along with the inductive case ("$\Theta(k)$ is true and it implies $\Theta(k+1)$ is true for all natural k") being true together implying that $\Theta(n)$ is true for any natural number n [5].

2. Generalizability of Function Property for Functionals on $\mathbb{R}^n$

In order to analyze the existence of the Inheritance and Generalizability properties, the procedure for constructing a model based on definitions 1.3 and 1.4 shall be demonstrated. As a convention, this model shall be called the Seyedi-Rohanin model (SRM):

Procedure 2.1 (SRM):

Categorize set $\mathcal{A}$ into all their subset groups via their names, i.e., let $\mathcal{B}_1$ be a set of all convex functions on $\mathbb{R}^n$, $\mathcal{B}_2$ be a set of all linear functions on $\mathbb{R}^n$, etc. Then index all these subset groups via indexed family (see Definition 1.3) i.e., f_{d_1} for the set $\mathcal{B}_1$, f_{d_2} for the set $\mathcal{B}_2$, etc. Then,

$$\alpha: f(d) = f_d,$$
$$\alpha: \mathbb{Z}_+ \longrightarrow \mathcal{A}$$

Then,

$d \neq c \rightarrow f_d \neq f_c \rightarrow f(d) \neq f(c)$ implies α is injective and corange(α)= $\bigcup_1^\infty \mathcal{B}_m$= $\mathcal{A}$ implies that α are surjective, $\forall f_d \in \mathcal{A} \Rightarrow f_d = f(d), \exists\, d \in \mathbb{N}$. Thus, α is bijective.

The sets $\mathcal{B}_m$ are countably infinite ($m = 1,2,\ldots$). $\mathcal{A} = \bigcup_1^\infty \mathcal{B}_m$ is countably infinite, because of $\bigcup_1^\infty \mathcal{B}_m$ is the union of all countably infinite sets (see Theorem 1.2). It implies that $\mathbb{N}$ and set $\mathcal{A}$ have a one-to-one correspondence. Thus, it implies that set $\mathcal{A}$ is countably infinite (see Theorem 1.1).

Let $\mathcal{Z}$ be a set of all functionals defined on $\mathbb{R}^n$. Similarly, set $\mathcal{Z}$ is categorized into all their subsets groups via their names, i.e., convex functionals, linear functionals, etc. Then all these groups are indexed.

Describe a map from set $\mathcal{A}$ to set $\mathcal{Z}$ to demonstrate their one-to-one correspondence. The following theorem shall establish the bijectivity of the map which was described between the set $\mathcal{A}$ and set $\mathcal{Z}$ in the model.

Theorem 2.1: Let the set $\mathcal{A}$ be the set of all functions defined on $\mathbb{R}^n$ and set $\mathcal{Z}$ be the set of all functionals defined on $\mathbb{R}^n$. If (α, β) represents the bijective maps from the natural number to both sets, then ψ as a map from set $\mathcal{A}$ to $\mathcal{Z}$is bijective.

Proof: Assuming that α and β are bijective it is implied that

$$\begin{cases} \alpha \text{ is injective} \\ \beta \text{ is injective} \end{cases} \Rightarrow$$

$$\begin{cases} i \neq e \rightarrow f_i \neq f_e \rightarrow: \mathrm{f}(i) \neq: f(e) \\ i \neq e \rightarrow g_i \neq g_e \rightarrow: g(i) \neq: g(e) \end{cases} \quad (2.1)$$

$$\text{and} \begin{cases} \alpha \text{ is surjective} \\ \beta \text{ is surjective} \end{cases} \Rightarrow$$

$$\begin{cases} \forall f_i \in \mathcal{A} \Rightarrow f_i = f(i), \exists\, i \in \mathbb{N} \\ \forall g_i \in \mathcal{Z} \Rightarrow g(i) = g_i, \exists\, e \in \mathbb{N} \end{cases} . \quad (2.2)$$

Based on (2.1), (2.2) and Definition 1.3:

$$\begin{cases} \psi: \mathcal{A} \longrightarrow \mathcal{Z}: i \neq e \rightarrow f_i \neq g_e \rightarrow f(i) \neq g(e) \\ \psi: \mathcal{A} \longrightarrow \mathcal{Z}: \forall g_i \in \mathcal{Z} \Rightarrow \exists f_i \in \mathcal{A}: f_i = g_i \end{cases}$$

$$\begin{cases} \Rightarrow \psi \text{ is injective} \\ \Rightarrow \psi \text{ is surjective} \end{cases}$$

Then, ψ is bijective (see Figure 2.1)■.

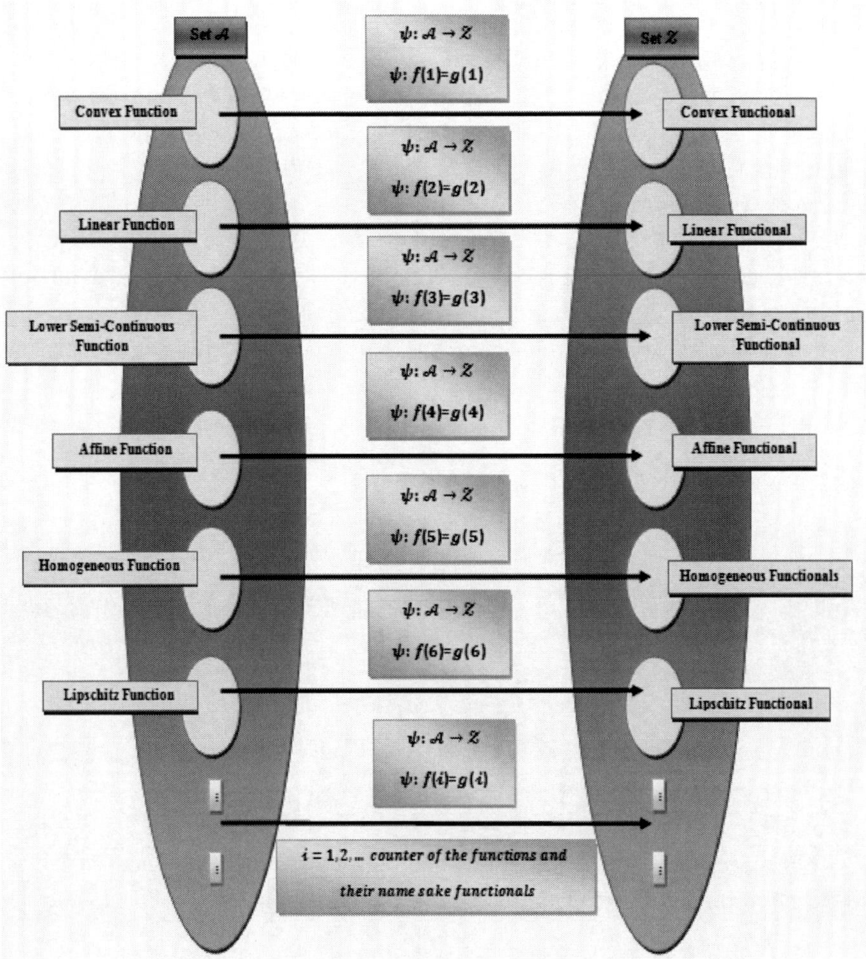

Figure 2.1. The existence of bijective maps from the set of all functions defined on $\mathbb{R}^n$ to the set of all functionals defined on $\mathbb{R}^n$ via SRM.

Based on Definition 1.4, let Θ be a predicate that performs the existence of the bijective maps $\psi_{k}: \mathcal{A} \longrightarrow \mathcal{Z}$, which connect the similar elements of the sets together (where $k \in \mathbb{N}$). Based on the construction of the sets $\mathcal{A}$ and $\mathcal{Z}$, it is clear when the sets $\mathcal{A}$ and $\mathcal{Z}$ are empty then $\Theta(0)$ is true, because it is also empty. And while the sets $\mathcal{A}$ and $\mathcal{Z}$ have one element i.e., $\mathcal{B}_1$ (set of all convex functions defined on $\mathbb{R}^n$) and $\mathcal{B}_1^*$ (set of all convex functionals defined on $\mathbb{R}^n$) then $\Theta(1)$ is true because there is a bijective map ψ_1 which connects both of them together based on Definition 1.3 and Theorem 2.1. Also while the sets $\mathcal{A}$ and $\mathcal{Z}$ have two elements i.e., $\mathcal{B}_1$ (set of all convex functions defined on $\mathbb{R}^n$), $\mathcal{B}_2$ (set of all linear functions defined on $\mathbb{R}^n$) and $\mathcal{B}_1^*$ (set of all convex functionals defined on $\mathbb{R}^n$), $\mathcal{B}_2^*$ (set of all linear functionals defined on $\mathbb{R}^n$) then $\Theta(2)$ is true because there are two bijective maps ψ_1 and ψ_2 which connect $\mathcal{B}_1$ to $\mathcal{B}_1^*$ and $\mathcal{B}_2$ to $\mathcal{B}_2^*$ respectively based on Definition 1.3 and Theorem 2.1. It means that $\Theta(2)$ is true because $\Theta(1)$ is true. As the process continues, it is also clear that $\Theta(3)$ is true because $\Theta(2)$ is true. Finally, $\Theta(k)$ is true because $\Theta(k-1)$ is true and also $\Theta(k+1)$ is true because $\Theta(k)$ is true. It induces that the conditions of Definitions 1.4 hold and all elements of the Set $\mathcal{A}$ is connected to their similar elements of the Set $\mathcal{Z}$ which have the same name i.e., $\mathcal{B}_{k}$ (affine functions on $\mathbb{R}^n$) and $\mathcal{B}_{k}^*$ (affine functionals on $\mathbb{R}^n$), (see Figure 2.1).

As a result of the above procedure, the following remark will appear.

Remark 2.1 (Generalizability Property): SRM shows that if there is a type of function defined on $\mathbb{R}^n$; then there is its same-name functional on $\mathbb{R}^n$, i.e., if there is a linear function defined on $\mathbb{R}^n$then there is a linear functional on $\mathbb{R}^n$ (see Figure 2.1).

3. Inheritance of Function Property in Functionals

In this section the Inheritance of function property for functional on $\mathbb{R}^n$ is described through the following theorem. Moreover, it is evident that $\mathcal{Z}$ (the set of all functionals defined on $\mathbb{R}^n$) is a subset of $\mathcal{A}$ (the set of all functions defined on $\mathbb{R}^n$). In other words, as it is defined when a functional is a type of function, then $\mathcal{Z} \subseteq \mathcal{A}$.

Theorem 3.1 (SUBSET PRINCIPLE): If $\Xi(x)$ is a condition on sets, then, for each set $\mathcal{A}$, there exists a unique set whose members are precisely those members $\mathcal{Z}$ of $\mathcal{A}$ for which $\Xi(\mathcal{Z})$ holds [6]. where $\Xi(.)$ is the Laws and Properties of functions.

Remark 3.1 (Inheritance Property): Based on Theorem 3.1, it is clear that subsets inherit their properties from their superset. It implies that any functional defined on $\mathbb{R}^n$, should observe all function laws and properties defined on $\mathbb{R}^n$ as its inherited property.

SRM, Theorem 2.1, Theorem 3.1 and bijectivity of the map ψ demonstrate that there exist the following results, such that they validate the existence of Inheritance and Generalizability properties from functions into functionals:

1. All functionals defined on $\mathbb{R}^n$, inherit their properties from their same-name functions on $\mathbb{R}^n$;
2. If there is a type of function defined on $\mathbb{R}^n$ then there is its same-name functional on $\mathbb{R}^n$, i.e., if there is a concave function defined on $\mathbb{R}^n$ then there is a concave functional on $\mathbb{R}^n$.

As a result of the above discussion, the SRM proposes the following theorem based on the Inheritance and Generalizibility properties in general.

Theorem 3.2: If and only if there is a definition for functions on $\mathbb{R}^n$ then there is its corresponding definition for functionals on $\mathbb{R}^n$.

Proof: Based on Procedure 2.1:

3. *Describe two sets $\mathcal{U}_1$ as the set of all function definitions on $\mathbb{R}^n$ which is countable and indexed via SRM and $\mathcal{U}_2$ as the set of all functional definitions on $\mathbb{R}^n$ which is countable and indexed via SRM. A map ψ which was introduced from the set $\mathcal{U}_1$ into $\mathcal{U}_2$ is bijective based on SRM. It implies that the sets $\mathcal{U}_1$ and $\mathcal{U}_2$ are one-to-one correspondence and the set $\mathcal{U}_2$ inherits its properties from its superset $\mathcal{U}_1$ (see Figure 3.1).*

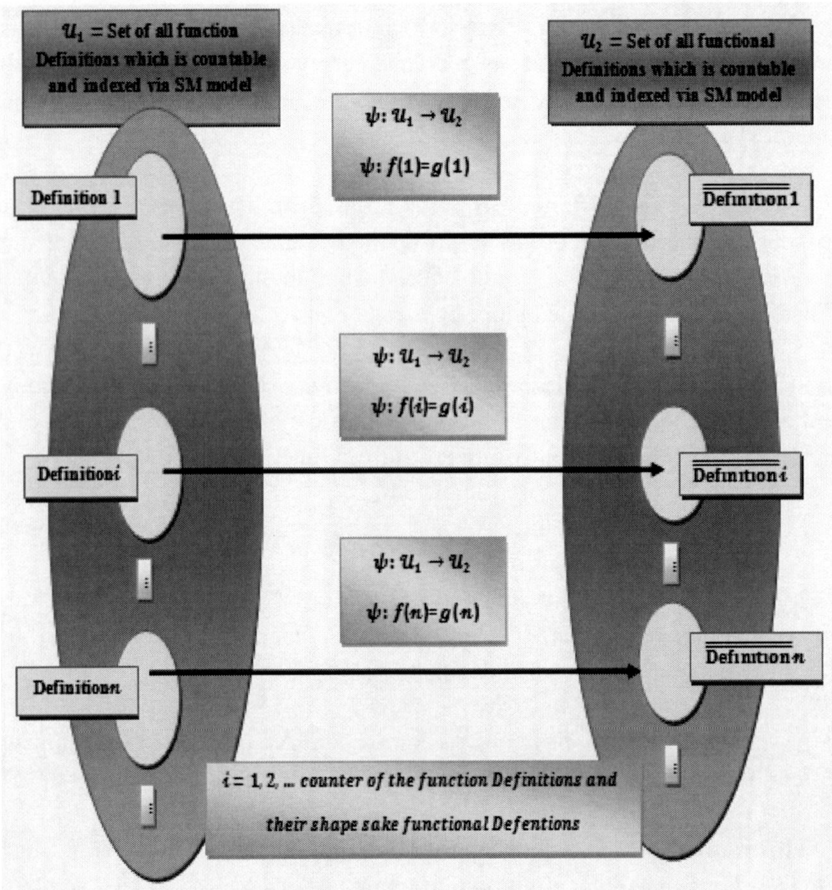

Figure 3.1. Existence of bijective maps from the set of all function definitions on $\mathbb{R}^n$ to the set of all functional definitions on $\mathbb{R}^n$ via SRM.

4. *Also, let* Θ *be the statement presenting the existence of the bijective maps* $\psi_{k}: \mathcal{U}_1 \longrightarrow \mathcal{U}_2$*, which connect the similar elements of the sets together (where* $k \in \mathbb{N}$*). Based on construction of the sets* $\mathcal{U}_1$ *and* $\mathcal{U}_2$*, it is clear when the sets* $\mathcal{U}_1$ *and* $\mathcal{U}_2$ *have one element, i.e.,* Δ_1 *(Abstract convexity definition for functions on* $\mathbb{R}^n$*) and* $\overline{\overline{\Delta_1}}$ *(Abstract convexity definition for functionals on* $\mathbb{R}^n$*) respectively, then* $\Theta(1)$ *is true because there is a bijective map* ψ_1 *which connects both of them together based on Theorem 2.1. Also while the sets* $\mathcal{U}_1$ *and* $\mathcal{U}_2$ *have two elements, i.e.,* Δ_1 *(Abstract convexity*

definition for functions on $\mathbb{R}^n$*),* Δ_2 *(Homogeneity definition for functions on* $\mathbb{R}^n$*) and* $\overline{\overline{\Delta_1}}$ *(Abstract convexity definition for functionals on* $\mathbb{R}^n$*),* $\overline{\overline{\Delta_2}}$ *(Homogeneity definition for functionals on* $\mathbb{R}^n$*) respectively, then* $\Theta(2)$ *is true because there are two bijective maps* ψ_1 *and* ψ_2 *which connect* Δ_1 *to* $\overline{\overline{\Delta_1}}$ *and* Δ_2 *to* $\overline{\overline{\Delta_2}}$ *respectively based on Theorem 2.1. It means that* $\Theta(2)$ *is true because* $\Theta(1)$ *is true. It is clearly induced that* $\Theta(3)$ *is true since* $\Theta(2)$ *is true, etc. Finally,* $\Theta(\mathcal{k})$ *is true because* $\Theta(\mathcal{k}-1)$ *is true and* $\Theta(\mathcal{k}+1)$ *is true because* $\Theta(\mathcal{k})$ *is true. It is undrestood that the condition of Definition 1.4 holds and all elements of the set* $\mathcal{U}_1$ *are connected to their similar elements of the set* $\mathcal{U}_2$ *which have the same name, i.e.,* $\Delta_{\mathcal{k}}$ *(increasing positive homogeneity definition for functions on* $\mathbb{R}^n$*) and* $\overline{\overline{\Delta_{\mathcal{k}}}}$ *(increasing positive homogeneity definition for functionals on* $\mathbb{R}^n$*).*

Inheritance and Generalizability properties and Theorem 3.2 claimed that all the definitions of functions defined on $\mathbb{R}^n$, can be extended to introduce new definitions for functionals on $\mathbb{R}^n$, i.e., increasing positive homogeneity definition for function on $\mathbb{R}^n$ can be extended to increasing positive homogeneity for functional on $\mathbb{R}^n$. The next section shall indicate the definition of Increasing Positively Homogeneous functional based on SRM in investigating the validity of the Theorem 3.2 experimentally.

4. Overview

In this part some classes of monotonic functionals either defined on the cone $\mathbb{R}^n_+ = \{x \in \mathbb{R}^n : x_i \geq 0 \; for \; all \; i = 1, \dots, n\}$ or defined on the cone $\mathbb{R}^n_{++} = \{x \in \mathbb{R}^n : x_i > 0 \; for \; all \; i = 1, \dots, n\}$, as well as the so-called normal subsets of these cones are going to be analysed. A nonempty set $\mathcal{U} \in \mathbb{R}^n_+$ is said to be normal if $(x \in \mathcal{U}, x' \in \mathbb{R}^n_+, x' \leq x) \Rightarrow x' \in \mathcal{U}$. In this paper, let the empty set be a normal subset of $\mathbb{R}^n_+$. Normal sets are connected to the so-called increasing positively homogeneous of degree one (IPH) functionals. An evident similarity between the class of IPH functionals and the class of sublinear functionals, and between normal sets and convex sets can be seen. Linear functionals are regarded as one of the major tools by which Sublinearity of functionals and the convexity of sets can be studied. For instance, a functional P defined on $\mathbb{R}^n$ is lower semicontinuous and sublinear if and only if this functional is abstract convex respecting the set of all linear functionals; on the other hand, there exists a set of linear functionals U such that

$$P(x) = \sup\{u(x) : u \in \mathcal{U}\}.$$

This section is going to put forwards some related definitions and properties on functions and fucntionals stated in [7-11].

Firstly, let $F: \tilde{Q} \to \mathbb{R}$ be a functional and $\tilde{Q}$ be real linear space.

1) F is subadditive if and only if:
   for all vectors x, y of $\tilde{Q}$ holds $F(x + y) \leq F(x) + F(y)$, [12].
2) F is additive if and only if:
   for all vectors x, y of $\tilde{Q}$ holds $F(x + y) = F(x) + F(y)$, [13].
3) F is homogeneous if and only if:
   for every vector x of $\tilde{Q}$ and for every real number b holds $F(b.x) = b.F(x)$, [14].
4) F is positively homogeneous if and only if:
   for every vector x of $\tilde{Q}$ and for every real number b such that $b > 0$ holds $F(b.x) = b.F(x)$, [15].
5) F is semi-homogeneous if and only if:
   for every vector x of $\tilde{Q}$ and for every real number b such that $b \geq 0$ holds $F(b.x) = b.F(x)$, [16].
6) F is absolutely homogeneous if and only if:
   for every vector x of $\tilde{Q}$ and for every real number b holds $F(b.x) = |b|.F(x)$, [17].

Secondly, let us consider $\tilde{Q}$. The following observations stated in [18], [19]:

A) every functional in $\tilde{Q}$ which is additive is also subadditive.
B) every functional in $\tilde{Q}$ which is homogeneous is also positively homogeneous.
C) every functional in $\tilde{Q}$ which is semi-homogeneous is also positively homogeneous.
D) A linear functional in $\tilde{Q}$ is an additive homogeneous functional in $\tilde{Q}$.

The following propositions were also stated in [9], [10]:

a. For every functional G in $\tilde{Q}$ and for every vector v of $\tilde{Q}$ holds $G(-v) = -G(v)$, [9].
b. For every functional G in $\tilde{Q}$ and for every vectors v_1, v_2 of $\tilde{Q}$ holds $G(v_1 - v_2) = G(v_1) - G(v_2)$, [9].
c. It is well-known that a positively homogeneous functional is convex if and only if it is subadditive, [10].
d. A positive homogeneous and subadditive function is called Sublinear, [10].

The next section shall present Increasing Positively Homogeneous (IPH) functionals based on the Inheritance and Generalizability of function on functionals.

5. Increasing Positively Homogeneous Functional Definition

Author of the reference [8] stated that the definition of IPH functional shall be proposed and extended.

Let $\tilde{Q}$ be either $\mathbb{R}^n_+$ or $\mathbb{R}^n_{++}$. A functional F defined on $\tilde{Q}$ is called increasing if $x \geq y$ implies that $F(x) \geq F(y)$. The functional F is positively homogeneous of degree one if $F(bx) = bF(x)$ for all $x \in \tilde{Q}$, $b > 0$ and $b \in \mathbb{R}_+$. If these definitions are combined the obtained result will be the following definition of an IPH functional

Definition 5.1: Let $\tilde{Q}$ be either $\mathbb{R}^n_+$ or $\mathbb{R}^n_{++}$. A functional $F: \tilde{Q} \to \mathbb{R}_{+\infty}$ is an IPH functional if the following conditions are satisfied:

1 $x \geq y$ implies that $F(x) \geq F(y)$:
2 $(bx) = bF(x)$ for all $x \in \tilde{Q}$ and $b > 0$ and $b \in \mathbb{R}_+$, [8].

Now some examples of IPH functionals are given.

1. *Let $\mathcal{R}$ be the set of all compact subsets of $\mathbb{R}^n$. Let $\mathcal{K}$ (respectively $\mathcal{K}_+$) denote the set of positive real-valued upper semicontinuous (respectively continuous) functions on $\mathbb{R}^n$ with compact support. And $c: \mathcal{R} \to \mathbb{R}_+$ is a positive and real-*

valued function defined on the compact subsets of $\mathbb{R}^n$. *Then, the Capacity functionals in the form* $C: h \to \int hdc$ *which were stated in [8] is IPH functionals where* $h \in \mathcal{K}_+$ *and* $\int hdc \in \mathbb{R}_+$.

2. Let $\widetilde{\mathcal{X}}$ be a topological space and subspace of $\mathbb{R}^n$. Let $G: \widetilde{\mathcal{X}} \to \mathbb{R} \cup \{+\infty\} \cup \{-\infty\}$ be a functional. Hamel (2004) defined the set $\mathcal{A}_G := \{x \in \widetilde{\mathcal{X}}: G(x) < 0\}$. Let $\mathcal{A} \subset \widetilde{\mathcal{X}}$ be a set and $k^0 \in \widetilde{\mathcal{X}}\{0\}$. Then the functional $\begin{cases} G_{\mathcal{A}}(x) = \inf\{t \in \mathbb{R}: x + tk^0 \in \mathcal{A}\} \\ \text{and } x_1 \leq x_2 \Rightarrow G_{\mathcal{A}}(x_1) \leq G_{\mathcal{A}}(x_2) \end{cases}$ which was stated in [10] is IPH functional.

6. Subdifferential of Increasing Positively Homogeneous Functionals

Two leading subdifferential definitions of functional and of IPH function respectively stated in [7], [11] are needed to be introduced and extended in order to explain the subdifferential of IPH functionals.

Let F be an IPH functional defined on $\mathbb{R}^n$, a local subdifferential of F at the vector $x_0 \in dom\, F$ $(dom\, F = \mathcal{U} \subseteq \mathbb{R}^n)$ is the set:

$$\partial F(x_0, \mathcal{U}) = \{v \in \mathbb{R}^n | \langle v, x - x_0 \rangle_{\mathbb{R}^n} \leq F(x) - F(x_0) \text{ for all } \mathcal{U} \in \mathbb{R}^n\}.$$

The above set is called the subdifferential of F at the vector x_0. As the process continues, the following definition of $\mathcal{L}$-subdifferential of IPH function stated in [7] for functional shall be extended.

Definition 6.1.: Let $\mathcal{L}$ be a set of simple functionals defined on a set $\mathcal{X}$. A functional $L \in \mathcal{L}$ is called an abstract subgradient (or $\mathcal{L}$-subgradient) of a proper functional $F: \mathcal{X} \to \mathbb{R}_{+\infty}$ at a vector y if $F(x) \geq L(x) - (L(y) - F(y))$ for all $x \in \mathcal{X}$. The set $\partial_{\mathcal{L}} F(y)$ of all abstract subgradients of F at y is referred to as the abstract subdifferential (or $\mathcal{L}$-subdifferential) of the functional F at the vector y.

Proposition 6.1: Let $\mathcal{L}$ be set of abstract linear functionals defined on a conic set $\mathcal{X}$ and let each $L \in \mathcal{L}$ be positively homogeneous of degree one $(L(bx) = bL(x)$ for all $b > 0)$. Then, for an $\mathcal{L}$-convex functional P and for $y \in \mathcal{X}$ thus:

$$\partial_{\mathcal{L}}P(y) = \{L \in supp(P,\mathcal{L}): L(y) = P(y)\} \tag{6.1}$$

Proof: like the proof on functions in [7], let $L \in \partial_{\mathcal{L}}P(y)$, that is,

$P(x) \geq L(x) - (L(y) - P(y))$ for every $x \in \mathcal{X}$. Let $x \in \mathcal{X}$. Investigate vectors bx with $b > 0$. Then

$$bP(x) = P(bx) \geq bL(x) - (L(y) - P(y)) \tag{6.2}$$

$$\Rightarrow bP(x) = L(x) - \frac{L(y)-P(y)}{b}.$$

Passing to the limit as $b \to +\infty$, $P(x) \geq L(x)$ is obtained, that is, $L \in supp(P,\mathcal{L})$. Specially, we have $L(y) \leq P(y)$. On the other hand, passing to the limit in (6.2) as $b \to +\infty$, $L(y) \geq P(y)$ will be attained. Therefore, $L(y) = P(y)$ so $\partial_{\mathcal{L}}P(y)$ is included in the set on the right-hand side in (6.1). Now assume that $P(x) \geq L(x)$ for all $x \in \mathcal{X}$ and $P(y) - L(y) = 0$. Then,

$P(x) \geq L(x) - (L(y) - P(y))$ for all $x \in \mathcal{X}$.

Consequently, the objective does not change. ■

As mentioned earlier, the $\mathcal{L}$-subdifferential of the functional P at a vector $y \in \mathbb{R}^n_{++}$ is given by

$$\partial_{\mathcal{L}}P(y) = \{L \in \mathcal{L}: P(x) \geq \langle L,x\rangle - \big(\langle L,y\rangle - P(y)\big) \text{ for all } x \in \mathbb{R}^n_{++}\}.$$

Since the set $\mathcal{L}$ consists of positively homogeneous functionals, it exploits from the above proposition that

$$\partial_{\mathcal{L}}P(y) = \{L \in supp(P,\mathcal{L}): \langle L,y\rangle = P(y)\}.$$

CONCLUSION

The study proved the existence of Inheritance and Generalizability of function properties on functionals. Furthermore, it was concluded this fact that if there is any type of function defined on $\mathbb{R}^n$ then there is its same-name functional on $\mathbb{R}^n$ is advocated by the Generalizability property. In addition, Inheritance property validates that the properties of function are inherited in its same-name functional and SRM is a major tool by which these properties are showed. It was also present in this paper that the extension of SRM based on indexed family and Axiom of Induction demonstrates the existence of the properties in definitions of both functionals and functions. Thus, the conclusion is that all the definitions of functions can be broadened to propose the new same definitions on functionals. What is more, this paper introduces the IPH functionals and their subdifferentiability based on these properties on functionals.

REFERENCES

[1] E. W. Weisstein, Functional Analysis, MathWorld – A Wolfram Web Resource, (1999). http://mathworld.wolfram.com/Functional Analysis. html.

[2] R. Todd, Functional, From MathWorld – A Wolfram Web Resource, created by Eric W. Weisstein, (1999), http://mathworld. wolfram.com/ Functional.html.

[3] S. Lang, Algebra, Addison-Wesley, (1993).

[4] J. R. Munkres, Topology, United States of America: Prentice Hall, Inc. upper saddle River, (2000).

[5] V. J. Katz, History of Mathematics: An Introduction, Addison-Wesley, (1998).

[6] M.O. Searcoid, Elements of Abstract Analysis, Springer undergraduate series, Berlin Heidelberg New York: Springer-Verlag, (1948).

[7] A.M. Rubinov, Abstract Convexity and Global Optimization. Nonconvex Optimization and its Applications, Kluwer Academic Publishers, (2000).

[8] B. Anger, Representation of Capacities, Matt. Ann., Springer-Verlag, 229, 245-258, (1977).

[9] B. Nowak, and A. Trybulec, Hahn-Banach Theorem, Journal of Formalized Mathematics, Volume 5, (2004).

[10] A. H. Hamel, From real-to set-valued coherent risk measures, Technical Report, Martin-Luther-University Halle-Wittenberg, (2004).

[11] P. O. Kasyanov, V. S. Mel'nik, and M. Piccirillo (2008). Local Subdifferentials and Multivariationals Inequalities in Banach and Frechet Spaces, Opuscula Mathematica, Vol. 28, No. 3.

[12] C. Byli´nski, Functions and their Basic Properties, Journal of Formalized Mathematics, 1(1): 223-225, (1989). http://mizar.org/ JFM/Vol1/funct_1.html.

[13] C. Byli´nski, Functions from a Set to a Set, Journal of Formalized Mathematics, 1(1): 153-155, (1989). http://mizar.org/JFM/Vol1/funct_2.html.
[14] C. Byli´nski, Partial Functions, Journal of Formalized Mathematics, 1(2), 357-359, (1989). http://www.mizar.org/JFM/Vol1/partfun1.html.
[15] C. Byli´nski, Some Basic Properties of Sets, Journal of Formalized Mathematics, 1, (1989d). http://www.mizar.org/JFM/Vol1/zfmisc_1.html.
[16] K. Hryniewiecki, Basic Properties of Real Numbers, Journal of Formalized Mathematics, 1, (1989).
[17] J. Kotowicz, and Y. Sakai, Properties of Partial Functions from a Domain to the Set of Real Numbers, Journal of Formalized Mathematics, 1, (1989).
[18] J. Popiołek, Some properties of functions modul and signum, Journal of Formalized Mathematics, 1, (1989). http://mizar.org/JFM/Vol1/absvalue.html.
[19] J. Popiołek, Real normed space, Journal of Formalized Mathematics, 2, (1990). http://mizar.org/JFM/Vol2/normsp_1.html.

In: A Note on Inheritance …
Editor: Iraj Sadegh Amiri
ISBN: 978-1-63485-784-0

Chapter 4

THE CUTTING ANGLE METHOD AND A LOCAL SEARCH ON OPTIMAL CONTROL PROBLEMS

Seyedalireza Seyedi[1,*], Iraj Sadegh Amiri[2,†], Sara Chaghervand[3,‡], Rohanin Ahmad[4,§] and Mohd Ismail abd Aziz[4,‖]

[1]Department of Mathematics, Hamedan Branch, Islamic Azad University, Hamedan, Iran
[2]Photonics Research Centre, University of Malaya, Kuala Lumpur, Malaysia
[3]Department of Electrical Engineering, Borujerd Branch, Islamic Azad University, Borujerd, Iran
[4]Department of Mathematics, Faculty Science, UniversitiTeknologi Malaysia, Skudai, Johor, Malaysia

---

* E-mail: seyedalireza.seyedi@gmail.com.
† Email: amiri@um.edu.my.
‡ E-mail: engsara1@yahoo.com.
§ E-mail: rohanin@utm.my,
‖ E-mismail@mel.fs.utm.my.

ABSTRACT

Optimal control is a study on way of choosing the best set of decisions to accomplish a particular objective. The paper mainly concerns the combinations of the Cutting Angle Method (CAM) with a local search in global optimization for solving functionals. Based on using the unit vector, a new technique is proposed to employ the CAM directly on optimal control problems. The numerical findings indicated that the proposed approach is very beneficial in the search for a global minimum.

Keywords: optimal control problem, optimization, cutting angle method, dynamic integrated system optimization and parameter estimation algorithm

1. INTRODUCTION

Optimal control is a study on how to select the best set of decisions to achieve an appropriate objective [1]. Furthermore, the performance of physical processes hinges on a great number of decisions. Physical processes can be divided into two parts including the static process and the dynamic process. The latter can be further divided into two, the linear and the non-linear process. The concern of this study is with the non-linear dynamic process.

Optimization of the non-linear dynamic process would model the process as a linear process. Then it is solved iteratively by using an appropriate dynamic optimization algorithm. Although there are some inaccuracies in mathematical models used, the solution of the linear process will converge to the solution of the original non-linear problem [2, 3].

Many algorithms are available for solving Optimal Control Problems (OCP). However, they are basically local search algorithms. This study shall propose the cutting angle method (CAM), a dynamic global search algorithm, which was introduced by [4] as a generalization of the cutting plane method from convex minimization for solving OCP. CAM has been successfully used on functions, however, OCP are functionals. The CAM can be employed in a wide class of non-convex global optimization problems [5]. The authors in [4] claimed CAM can help local search algorithms to converge to global optima. This method is used in combination with a local search algorithm to arrive at a true global solution [4]. The approach basically begins with a solution of a local search algorithm. Then, the cutting angle method is applied and repeated until one arrives at a solution that is lower than the current solution of the local search algorithm. Using the lower solution as the new initial solution, the local search algorithm is again run on the problem. Once a new solution is obtained from this local search, the cutting angle method is once again applied. The process would stop only when the cutting angle method fails to find a lower valued solution after a predetermined number of iterations. The new solution of the local search algorithm would be regarded as the global optimum.

As a case study, this paper proposes to use a local search algorithm in combination with the CAM. In this study the Dynamic Integrated Systems Optimization and Parameter Estimation (DISOPE) algorithm have been chosen as the local search. The DISOPE algorithm in this research is an extension of an earlier algorithm called Integrated System Optimization and Parameter Estimation (ISOPE). It was originally developed by [6, 7] for on-line steady state optimization of industrial processes implemented through adjustment of regulatory controller set points. This algorithm has been substantiated to be successful in solving a great deal of example problems [7, 8, 9, 10]. Later, authors in [11] elicited sufficient conditions for convergence of this algorithm. A fundamental feature of ISOPE is that the iterations converged to the correct real optimum. ISOPE was planned for the steady-state optimizing control. Naturally, it was later developed to solve the dynamical optimal control problems. Author in [12] extended ISOPE to dynamical problems and it has been termed very much the same. As it was originally developed and published, DISOPE addressed continuous-time, unconstrained, centralized and time invariant optimal control problems [3]. Author in [3] developed and improved the existing knowledge on the technique in order to make it agreeable for its implementation in the process industry.

DISOPE was initially developed for continuous nonlinear optimal control problems [2] and then extended to discrete systems [13] and to optimal tracking control problems [14]. The technique has also been extended to deal with un-matched terminal constraints [15]. A range of application of DISOPE techniques has also been developed by [3]. Considering everything, this algorithm is a type of an algorithm which was particularly written to solve OCP using the basic gradient as search direction. The CAM employed special transformed objective function for each intermediate use of the method. Authors in [16] applied real valued functions for the objective functions in their solved problems. This study mainly focuses on proposing a technique to use CAM for solving OCP directly through a combination with a local search. The CAM was combined successfully with the Discrete Gradient Method

(DGM) on functions by [16]. The current research introduces the combination of CAM and DISOPE algorithm on OCP.

As it was said before, Authors in [4] proposed the cutting angle method which is able to help local search algorithms to converge to global optimum. This method is used in combination with a local search algorithm to arrive at a true global solution [4]. This section concerns a combination of CAM and a local search in global optimization. Unit Vector Combinations Technique (UVCT) is proposed to use of CAM directly for OCP. The reported numerical results demonstrate that the proposed approach is very useful in the search for a global optimization. Additionally, the finding of this section shall verify that the output of the DISOPE algorithm can be truly the initial solution to CAM. Moreover, based on the result of the combination, CAM confirms that the initial solution given by DISOPE is actually already the global solution. The combinations of the CAM in global optimization and DGM as a local search were put forward by [16] which successfully substantiated their claim about CAM.

In particular, DGM is a type of gradient descent method by which CAM could help overcome the above-mentioned problems and help reach the exact solution via suitable choices of the support functions. Based on the ability of CAM in global optimization and its earlier combination with DGM, and the property of DISOPE algorithm which is a type of gradient descent method, the combination of CAM and DISOPE algorithm is proposed as follows and is able to arrive at the global solution.

2. A Combination of Cutting Angle Method and a Local Search on Optimal Control Problems

The subsequent procedure shall analyze an application of CAM on OCP based on discretization techniques as well as the Inheritance and Generalizability of function optimization algorithms to solve functional (in Particular OCP) which introduced by [17]

Procedure 1:

I. (Discretization): The original problem in this research is in the form of:

$$\text{Minimize } J = \int_{t_0}^{t_1} E(x,u,t)dt \tag{1}$$

$$s.t\ \dot{x} = g(x,u,t) \tag{2}$$
$$x(t_0) = x_0,\ x(t_1) = x_1,$$

where $E: \mathcal{A} \times \mathcal{B} \times [t_0, t_1] \to \mathbb{R}$ is a continuous function, g is a continuous nonlinear time varying function, $g: \mathcal{A} \times \mathcal{B} \times [t_0, t_1] \to \mathbb{R}^n, t \in [t_0, t_1] \subseteq \mathbb{R}$, , $x(t) \in \mathcal{A}$, $u(t) \in \mathcal{B}$ and $\mathcal{A} \subseteq \mathbb{R}^n, \mathcal{B} \subseteq \mathbb{R}^m$ are compact subset and must be chosen so that the system reach from initial state $x(t_0)$ to final state $x(t_1)$.

Firstly, discretize the original problem (1) to the following equivalent problem and partition intervals $[t_0, t_1]$ to n parts and also $t \in [t_0, t_1]$.

$$\text{Minimize } J = \mathfrak{h} \sum_{i=1}^{n-1} E_i\,(x,u,t) \tag{3}$$

$$s.t\ \mathrm{x} = \mathfrak{h} * g(x,u,t) \tag{4}$$
$$x(t_0) = x_0,,\ x(t_1) = x_1,$$

where $\mathfrak{h}$ is a step size of the discretization, $\mathfrak{h} = t_1 - t_0/n$, $x = \{x_j\}, u = \{u_k\}, j = 1,2,\dots,m_1$ and $k = 1,2,\dots,m_2$.

II. Secondly, apply CAM for solving the discretized OCP (3) via unit vectors Combinations technique (UVCT). This research generalizes CAM to solve discretized OCP via the UVCT as follows,

In Step 0 of CAM initialize the state vector x_1 by unit vectors

$$e^1 = (1,0,0,\dots,0,0)_{1\times n-1}, e^2 = (0,1,0,\dots,0,0)_{1\times n-1}, \dots,$$

$$e^{n-2} = (0,0,0,\dots,1,0)_{1\times n-1}, e^{n-1} = (0,0,0,\dots,0,1)_{1\times n-1}.$$

Next, calculate the corresponding values of $x_2, x_3, \dots, x_{m_1}; u_1, u_2, \dots, u_{m_2}$ via (4) and calculate the corresponding value of J_1 via (3). Then, construct basis vectors

$$L^p = \left(\frac{J(e^p, \mathrm{u}, t)}{e^p}\right), p = 1, \dots, n-1.$$

And apply CAM to find x_1^* based on support vectors $L^{k} = \left(\frac{J(x^*,u,t)}{x^*}\right), k =$ "the number of iterations of CAM". Then calculate the corresponding values of other components via (4) and J_1^*.

Next, continue this operation for the state vectors $x_2, x_3, \ldots, x_{m_1}$ to find $x_2^*, x_3^*, \ldots, x_{m_1}^*$ and corresponding optimal values $J_2^*, J_3^*, \ldots, J_{m_1}^*$.

The UVCT initializes 2, 3 and $m_1 - 1$ sate vectors combinations respectively by unit vectors and use CAM to find corresponding optimal values of them in parallel as follows.

CAM can be applied for one state vector or combination of two, three or more states vectors:

$$x_1, x_2; x_1, x_3; \ldots; x_1, x_{m_1}; x_2, x_1; x_2, x_3; \ldots;$$
$$x_2, x_{m_1}, \ldots x_1, x_2, x_3; \ldots; x_1, x_2, x_{m_1};$$
$$\ldots; x_{m_1}, x_{m_1-1}, \ldots, x_2$$

and find the corresponding optimal solutions

$$J_{12}^*, J_{13}^*, \ldots, J_{1m_1}^*; J_{21}^*, J_{23}^*, \ldots, J_{2m_1}^*;$$
$$\ldots J_{123}^*, \ldots, J_{12m_1}^*; \ldots; J_{m_1 m_1-1\ldots2}^*$$

respectively.

Similarly, CAM can be applied for one control vector or combination of two, three or more control vectors:

$$u_1, u_2, u_3, \ldots, u_{m_1}; u_1, u_2; u_1, u_3; \ldots; u_1, u_{m_1}; u_2, u_1; u_2, u_3;$$
$$\ldots; u_2, u_{m_1}, \ldots; u_1, u_2, u_3; \ldots; u_1, u_2, u_{m_1}; \ldots; \ldots.; u_{m_1}, u_{m_1-1}, \ldots, u_2$$

and find the corresponding optimal solutions

$$\overline{J_1}^*, \overline{J_2}^*, \overline{J_3}^*, \ldots, \overline{J_{m_1}}^*; \overline{J_{12}}^*, \overline{J_{13}}^*, \ldots, \overline{J_{1m_1}}^*; \overline{J_{21}}^*, \overline{J_{23}}^*, \ldots, \overline{J_{2m_1}}^*;$$
$$\ldots; \overline{J_{123}}^*, \ldots, \overline{J_{12m_1}}^*; \ldots; \overline{J_{m_1 m_1-1\ldots2}}^*.$$

Regarding the minimization of the discretized OCP by CAM the set

$$\mathcal{J}^* = \{J_1^*; J_2^*, J_3^*, \ldots, J_{m_1}^*; J_{12}^*, J_{13}^*, \ldots, J_{1m_1}^*; J_{21}^*, J_{23}^*, \ldots, J_{2m_1}^*; \ldots;$$
$$J_{123}^*, \ldots, J_{12m_1}^*; \ldots; \ldots; J_{m_1 m_1-1\ldots2}^*; \overline{J_1}^*, \overline{J_2}^*, \overline{J_3}^*, \ldots, \overline{J_{m_1}}^*;$$

$\overline{J_{12}}^*, \overline{J_{13}}^*, \dots, \overline{J_{1m_1}}^*; \overline{J_{21}}^*, \overline{J_{23}}^*, \dots, \overline{J_{2m_1}}^*; \dots; \overline{J_{123}}^*, \dots, \overline{J_{12m_1}}^*; \dots;$
$\dots; \overline{J_{m_1 m_1-1\dots2}}^*\}.$

will be obtained and the smallest elements of the set $\mathcal{J}^*$ will be the global optimum solution of the mentioned OCP based on the global property of CAM.

In some cases the following suggestions are helpful to decrease the processing time of the algorithm based on OCP assumed conditions.

i. In some OCP, just use the combinations of the vectors which involve into the performance index
ii. In some examples, just use state vectors or control vectors combinations to converge to the global optimum.

3. Generalization of Cutting Angle Method for Optimal Control Problems

Consider the following global optimization problem:

$$J(x, u, t) \to min$$

subject to $x(t), u(t) \in \mathcal{S}^*$,

where $\mathcal{S}^* = \{t \in \mathbb{R}^n_+ : \sum_i^r x_i(t) = 1, \sum_i^r u_i(t) = 1\}\}$ and r is the number of coordinates of the state and control vectors $x(t), u(t)$.

Let $G: \mathcal{S}^* \to \mathbb{R}$ be a positive Lipschitz functional defined on the unit simplex. Then it can be extended to a finite IPH functional $J(x, u, t)$ on the cone $\mathbb{R}^n_+$, which would coincide with $G(x, u, t) + c$ on $\mathcal{S}^*$. I.e., $J(x, u, t) = G(x, u, t) + c$ is an IPH functional on $\mathcal{S}^*$. Then, for each $x(t), u(t) \in \mathbb{R}^n_+$ define the support vectors

$$\begin{cases} L_x = L = \left(\frac{J(x,u,t)}{x}\right) = \left(\frac{J(x,u,t)}{x_1}, \frac{J(x,u,t)}{x_2}, \dots, \frac{J(x,u,t)}{x_n}\right) \\ L_u = \bar{\bar{L}} = \left(\frac{J(x,u,t)}{u}\right) = \left(\frac{J(x,u,t)}{u_1}, \frac{J(x,u,t)}{u_2}, \dots, \frac{J(x,u,t)}{u_n}\right) \end{cases}.$$

The n vectors $e^m = (0, \dots, 0, 1, 0, \dots, 0)$, with 1 in the mth position shall be used, and the corresponding support vectors $\begin{cases} L^m = \left(\frac{J(e^m, u, t)}{e^m}\right) \\ \overline{\overline{L}}^m = \left(\frac{J(x, e^m, t)}{e^m}\right) \end{cases}$, $m = 1, \dots, n$ shall be called basis vectors. A set of $k \geq n$ support vectors (and hence k known values of the functional $J(x, u, t)$ at k distinct vectors),

$$\begin{cases} \mathcal{K}_x = \{L^k\}_{j=1}^{k} \\ \mathcal{K}_u = \{\overline{\overline{L}}^k\}_{j=1}^{k} \end{cases}.$$

Let also the first n support vectors be the basis vectors. This choice of support vectors guarantees that the algorithm will locate all local (and hence global) minimizers of the auxiliary functionals

$$\begin{cases} H_k(x, u, t) = \max_{j \geq k} \min_{i=1,\dots,n} L_i^j x_i \\ \overline{\overline{H}}_k(x, u, t) = \max_{j \geq k} \min_{i=1,\dots,n} \overline{\overline{L}}_i^j u_i \end{cases}$$

in the interior of the unit simplex. It always underestimates the value of

$$J(x, u, t) \colon \begin{cases} H_k(x) \leq J(x, u, t) \\ \overline{\overline{H}}_k(x) \leq J(x, u, t) \end{cases}.$$

Hence,

$$\begin{cases} b_k = \min_{x \in S^*} H_k(x, u, t) \leq \min_{x \in S^*} J(x, u, t) \\ \overline{\overline{b}}_k = \min_{x \in S^*} \overline{\overline{H}}_k(x, u, t) \leq \min_{u \in S^*} J(x, u, t) \end{cases}.$$

On the other hand, the sequence of its minima, $\begin{cases} \{b_k\}_{k=n}^{\infty} \\ \{\overline{\overline{b}}_k\}_{k=n}^{\infty} \end{cases}$ is increasing, and convergence to the global minimum of $J(x, u, t)$ based on its function framework which was presented by [18, 4]. And based on function the CAM was formulated by [18]. The following algorithms are presented for functional in general (or for OCP in particular).

The CAM for OCP

Algorithm 1

Step (0). (Initialization)

(a) Take points e^m, $m = 1, \dots, n$, and construct basic vectors $L^m = \left(\frac{J(e^m,u,t)}{e^m}\right)$, $m = 1, \dots, n$.

(b) Define the functional $H_n(x) = \max_{j \le n} \min_{i=1,\dots,n} L_i^j x_i = \max_{j \le n} L_j^j x_j$.

(c) Set $k = n$.

Step 1. Find $x^* = \arg[\min_{x \in S^*} H_k(x)]$.

Step 2. Set $k = k + 1$ and $x^k = x^*$.

Step 3. Compute $L^j = \left(\frac{J(x^k,u,t)}{x^k}\right)$. Define the function

$$H_k(x) = \max_{j \le k} \min_{i=1,\dots,n} L_i^j x_i = \max\{H_{k-1}(x), \min_{i=1,\dots,n} L_i^k x_i\}.$$

Go to Step 2.

Step 1, minimization of the auxiliary functional is the most time consuming step of the algorithms 1 and 2. The next theorem will improve this problem.

Theorem 1: Let $x^*(t) > 0$ (or $u^*(t) > 0$) be a local minimizer of $H_k(x, u, t)$ (or $\bar{\bar{H}}_k(x, u, t)$) over the relative interior of S^*, $riS^* = \{x(t) \in S^*, x(t) > 0\}$ (or $riS^* = \{u(t) \in S^*, u(t) > 0\}$). Then there exists a subset $\mathcal{L} = \{L^{j_1}, L^{j_2}, \dots, L^{j_n}\}$ (or $\bar{\bar{\mathcal{L}}} = \{\bar{\bar{L}}^{j_1}, \bar{\bar{L}}^{j_2}, \dots, \bar{\bar{L}}^{j_n}\}$) of the set $\mathcal{K}$ (or $\bar{\bar{\mathcal{K}}}$), such that

(1) $x = \{d/L_1^{j_1}, d/L_2^{j_2}, \dots, d/L_n^{j_n}\}$ with $d = (\sum_i 1/L^{j_i})^{-1}$
(or $u = \{d/\bar{\bar{L}}_1^{j_1}, d/\bar{\bar{L}}_2^{j_2}, \dots, d/\bar{\bar{L}}_n^{j_n}\}$ with $d = (\sum_i 1/\bar{\bar{L}}^{j_i})^{-1}$).

(2) $\max_{j \le k} \min_{i=1,\dots,n} L_i^j / L_i^{j_i} = 1$
(or $\max_{j \le k} \min_{i=1,\dots,n} \bar{\bar{L}}_i^j / \bar{\bar{L}}_i^{j_i} = 1$.

(3) *Either* $\forall i: j_i = i$, *or* $\exists m: k_m > n, L_i^{j_m} > L_i^{j_i}, \forall i \ne m$
(or Either $\forall i: j_i = i$, or $\exists m: k_m > n, \bar{\bar{L}}_i^{j_m} > \bar{\bar{L}}_i^{j_i}, \forall i \ne m$.).

Proof: Completely similar to the proof on function in [16].

The value of the auxiliary functional at x (or u) is $H_{k}(x,u,t) = d$ (or $\bar{\bar{H}}_{k}(x,u,t) = d$).

The following version of CAM for functionals works based on the above theorem, by examining all possible combinations of n support vectors.

Consider a set of k support vectors $\mathcal{K} = \{L^{j}\}_{j=1}^{k}$ (or $\bar{\bar{\mathcal{K}}} = \{\bar{\bar{L}}^{j}\}_{j=1}^{k}$), $L^{j} \in \mathbb{R}_{+}^{n}$ (or $\bar{\bar{L}}^{j} \in \mathbb{R}_{+}^{n}$). Let $\mathcal{J}$ denote $\{1,2,\dots,n\}$. From Theorem 7.1, the local minima of the auxiliary functional $H_{k}(x,u,t)$ (or $\bar{\bar{H}}_{k}(x,u,t)$ are combinations of n support vectors $\mathcal{L} = \{L^{j_1}, L^{j_2}, \dots, L^{j_n}\}$ (or $\bar{\bar{\mathcal{L}}} = \{\bar{\bar{L}}^{j_1}, \bar{\bar{L}}^{j_2}, \dots, \bar{\bar{L}}^{j_n}\}$) that satisfy the following conditions:

(I) $\forall i, s \in \mathcal{J}, i \neq s$: $L_{i}^{j_i} < L_{i}^{j_s}$ (or $\bar{\bar{L}}_{i}^{j_i} < \bar{\bar{L}}_{i}^{j_s}$)

(II) $\forall v \in \mathcal{K} \backslash \mathcal{L}, \exists i \in \mathcal{J}$: $L_{i}^{j_i} \geq v_i$ (or $\forall \bar{\bar{v}} \in \bar{\bar{\mathcal{K}}} \backslash \bar{\bar{\mathcal{L}}}, \exists i \in \mathcal{J}$: $\bar{\bar{L}}_{i}^{j_i} \geq \bar{\bar{v}}_i$).

The subset $\mathcal{L}$ (or $\bar{\bar{\mathcal{L}}}$), which satisfied conditions (I) and (II) above, is called a valid combination of support vectors.

Let $\mathcal{W}^{k}$ (or $\bar{\bar{\mathcal{W}}}^{k}$) denote the set of all valid combinations $\mathcal{L}$ (or $\bar{\bar{\mathcal{L}}}$) of k support vectors satisfying conditions (I) and (II):

$$\mathcal{W}^{k} = \{\mathcal{L} = \{L^{j_1}, L^{j_2}, \dots, L^{j_n}\}, L^{j_i} \in \mathcal{K}: (1)\forall i, s \in \mathcal{J}, i \neq s: \mathcal{L}_{ii} < \mathcal{L}_{si}$$

and

$$\text{(II) } \forall v \in \mathcal{K} \backslash \mathcal{L}\ \exists i \in \mathcal{J}: \mathcal{L}_{ii} \geq v_i\}$$

$$\text{(or } \bar{\bar{\mathcal{W}}}^{k} = \{\bar{\bar{\mathcal{L}}} = \{\bar{\bar{L}}^{j_1}, \bar{\bar{L}}^{j_2}, \dots, \bar{\bar{L}}^{j_n}\}, \bar{\bar{L}}^{j_i} \in \bar{\bar{\mathcal{K}}}: (1)\forall i, s \in \mathcal{J},$$

$$i \neq s: \bar{\bar{\mathcal{L}}}_{ii} < \bar{\bar{\mathcal{L}}}_{si} \text{ and } (2)\ \forall \bar{\bar{v}} \in \bar{\bar{\mathcal{K}}} \backslash \bar{\bar{\mathcal{L}}}\ \exists i \in \mathcal{J}: \bar{\bar{\mathcal{L}}}_{ii} \geq \bar{\bar{v}}_i\})$$

The problem of finding local minima of $H_{k}(x,u,t)$ (or $\bar{\bar{H}}_{k}(x,u,t)$ is translated into the problem of listing the elements of $\mathcal{W}^{k}$ (or $\bar{\bar{\mathcal{W}}}^{k}$). A simplistic approach is then:

Step 1. Construct all combinations $\mathcal{L}$ (or $\bar{\bar{\mathcal{L}}}$) satisfying (I)

Step 2. Check the obtained combinations against (II).

Then improve on this by considering that all elements of $\mathcal{W}^k$ (or $\bar{\bar{\mathcal{W}}}^k$) that do not involve L^j (or $\bar{\bar{L}}^j$), (i.e., minima of $H_{k-1}(x,u,t)$ (or $\bar{\bar{H}}_{k-1}(x,u,t)$) do not need to be recomputed, hence the algorithm takes the form:

Algorithm 2

Step (0). (Initialization)

(a) Take points e^m, $m = 1, \dots, n$, and construct basic vectors $L^m = \left(\frac{J(e^m,u,t)}{e^m}\right)$, $m = 1, \dots, n$.

(b) Define the function $H_n(x,u,t) = \max_{j \le n} \min_{i=1,\dots,n} L_i^j x_i = \max_{j \le n} L_j^j x_j$.

(c) Set $k = n$. Set $\mathcal{W}^k = \{\{L^1, L^2, \dots, L^n\}\}$.

(d) Calculate $d = (\sum_{i=1,\dots n} 1/L_i^i)^{-1}$.

Step 1

(a) Retrieve all valid combinations $\mathcal{L}$ (i.e., $\mathcal{W}^k$).

(b) Select $\mathcal{L} \in \mathcal{W}^k$.

Step 2

(a) $k = k + 1$.

(b) Take $x^* = \frac{d}{\text{diag}}(\mathcal{L})$ and evaluate $J(x^*,u,t)$.

(c) Compute $L^k = (J(x^*,u,t)/x^*)$. Define the function

$$H_k(x,u,t) = \max_{j \le k} \min_{i=1,\dots,n} L_i^j x_i = \max\{H_{k-1}(x), \min_{i=1,\dots,n} L_i^k x_i\}$$

Step 3

(a) Check $\mathcal{W}^{k-1}$ against (II) and remove those that fail (II).

(b) Move the remaining combinations into $\mathcal{W}^k$.

Step 4

(a) Construct all combinations $\mathcal{L}$ that involve L^k and satisfy (I).

(b) Calculate $d = (\sum_{i=1,\ldots n} 1/L_i^{k_i})^{-1}$ for each such combination.
(c) Add these combinations to $\mathcal{W}^k$.
(d) Go to Step 1.

Remark 1: In Algorithm 2 if the diagonal of $\mathcal{L}$ (or $\bar{\bar{\mathcal{L}}}$) was dominated by L^k, this combination will be not moved into $\mathcal{W}^{k+1}$.

Remark 2: In Algorithm 2, each iteration the algorithm computes and checks the value of $J(x^*, u, t) - d$ (or $J(x, u^*, t) - d$) with the value of assumed stopping criteria. While it satisfies then the algorithm stopped.

Based on the combination of CAM and a local search (DGM) in [16], from (3) assume that

$$J = \mathfrak{h} \sum_{i=1}^{n-1} E_i\,(x, u, t) = g(x, u, t)$$

Then the following problem of global optimization is studied,

$$g(x, u, t) \to \min \text{ subject to } (x, u, t) \in \mathcal{S} \quad (5)$$

where

$$\mathcal{S} = \{x(t) = (x_1, \ldots, x_m) \in \mathbb{R}^n : \textstyle\sum_1^m x_i \; = \; 1, x_1 \geq 0, \ldots, x_m \geq 0;$$

$$u(t) = (u_1, \ldots, u_m) \in \mathbb{R}^n : \textstyle\sum_1^m u_i \; = \; 1, u_1 \geq 0, \ldots, u_m \geq 0\}$$

is the unit simplex and g is a ICAR function defined on $\mathcal{S}$ (unit simplex).

Applying the CAM to the minimization of the function g is not able to consider the known value of this function at stationary vector y found by DISOPE algorithm as a local search. In order to implement this value, a function g shall be transformed into a new function p. A function p is called a transformed function of g (with respect to vector y) if

1) $p(x, u, t) \leq g(x, u, t)$;
2) $\min\limits_{x,u \in \mathcal{S}} p(x, u, t) = \min\limits_{x,u \in \mathcal{S}} g(x, u, t)$. (6)

It follows from (6) that the set $\mathcal{T}(x^*, u^*, t) = \{x, u \in \mathcal{S}: g(x, u, t) \leq p(x^*, u^*, t)\}$ is nonempty for all x^*, u^*. Indeed if $(\overline{x^*}, \overline{u^*}, t)$ is a global minimizer of g over $\mathcal{S}$ then $g(\overline{x^*}, \overline{u^*}, t) = \min_{x,u \in \mathcal{S}} p(x, u, t) \leq p(x^*, u^*, t)$ so $(\overline{x^*}, \overline{u^*}, t) \in \mathcal{T}(x^*, u^*, t)$.

Proposition 2: The following functions are transformed for a function g regarding (y, u, t), [16]:

1) $p_1(x, u, t) = \min(g(x, u, t), g(y, u, t))$;
2) $p_2(x, u, t) = \min_{j = 1,..,n} \min_{b \in \mathcal{A}_j} p_1(b(x, u, t) + (1 - b)(x, u, t)^j)$ where $1 \in \mathcal{A}_j \subset [0,1]$, $(x, u, t)^j \in \mathcal{S}, j = 1, .., n$;
3) $p_3(x, u, t) = \min_{j = 1,..,n} \min_{b \in \mathcal{A}_j} g(b(x, u, t) + (1 - b)(x, u, t)^j)$ where $\mathcal{A}$ and $(x, u, t)^j$ are as in item 2);
4) $p_4(x, u, t) = \min_{b \in B} \min_{a \in \mathcal{A}} \sum_{j \in \mathcal{J}} b_j \, g(b(x, u, t) + (1 - b)(x, u, t)^j)$, where $\mathcal{J}$ is a finite set of indices, $\mathcal{B} = \left\{ (b_j)_{j \in \mathcal{J}} : \sum_{j \in \mathcal{J}} b_j = 1, b_j \geq 0 \ (j \in \mathcal{J}) \right\}$, $1 \in \mathcal{A}_j \subset [0,1]$, $(x, u, t)^j \in \mathcal{S} \ (j \in \mathcal{J})$;
5) $p_5(x, u, t) = \min_{b \in B} \min_{a \in \mathcal{A}} \sum_{j \in \mathcal{J}} b_j \, p_1(x, u, t)(b(x, u, t) + (1 - b)(x, u, t)^j)$, where $\mathcal{J}, \mathcal{B}, \mathcal{A}$ are as in item 4);
6) $p_6(x, u, t) = b p_1(x, u, t) + (1 - b) g(x, u, t)$ where $b \in (0,1)$.

What is more, the choice of the sets $\mathcal{A}_j$, $j = 1, .., n$, depends on the problem under consideration and in particular, on the number of vectors. On the one hand, the number of elements of $\mathcal{A}_j$ should be large enough in order to obtain a good minorant for the objective function g. On the other hand, it should not be too large; otherwise a large number of objective function evaluations at each iteration of the cutting angle method shall appear.

Remark 3:

a) Functions $p_2 - p_5$ depend on many parameters.
b) All functions $p_1 - p_6$ are nonsmooth [16].

Consider the problem of global optimization presented by (5). An algorithm is proposed for its solution. This algorithm is based on a combination of the CAM and a local search and a choice of transformed function p proposed in [16].

The Algorithm (CAM-DISOPE)

Step 0: (Initialization) Choose an arbitrary starting vectors $\overline{(x,u,t)}^0$. Set $i = 0$.

Step 1: Find stationary vectors of g over $\mathcal{S}$ by the DISOPE algorithm, starting from the vectors $(\bar{x},\bar{u},t)^i$. Denote this stationary vectors by $(\bar{x},\bar{u},t)^i$ and let $g_i = g((\bar{x},\bar{u},t)^i)$.

Step 2: Construct a transformed function p_i of the function g with respect to the vector $(x,u,t)^i$.

Step 3: Take vectors $(x,u,t)^r = e^r$, $r = 1,..,n$, $(x,u,t)^{n+1} = (x,u,t)^i$. Let $l^r = p^i((x,u,t)^r)/(x,u,t)^r$, $r = 1,..,n+1$ and set $j = n+1$. Construct the function

$$h_j\big((x,u,t)\big) = \max_{r\le j} \min_{i\in \mathcal{I}(l^r)} l_i^r (x,u,t)_i,$$
$$l^r = (l_1^r, \dots, l_n^r) \in \mathbb{R}^n, l_i^r \ge 0, i = 1, \dots, n, \mathcal{I}(l^r) = \{i: l_i^r > 0\},$$
$$r = 1, \dots, j, j \ge n.$$

Step 4: Solve the problem

$$h_j\big((x,u,t)\big) \to \min \text{ subject to } (x,u,t) \in \mathcal{S}. \tag{7}$$

Step 5: Let (x^*,u^*,t) be a solution of the problem (8.3). Set $j = j+1$ and $(x,u,t)^j = (x^*,u^*,t)$.

Step 6: Compute $p^* = p^i((x^*,u^*,t))$. If $p^* < g_i$ then find a vector $(x,u,t)' \in \mathcal{T}((x^*,u^*,t))$, set $i = i+1$, $(\bar{x},\bar{u},t)^i = (x,u,t)'$ and go step 1.

Step 7: Otherwise compute $l^j = p^i((x,u,t)^j)/(x,u,t)^j$, define the function

$$h_j(x,u,t) \to \max\{h_{j-1}(x,u,t) \to, \min_{i\in \mathcal{I}(l^j)} l_i^j(x,u,t_i)\} \equiv$$
$$\max_{r\le j} \min_{i\in \mathcal{I}(l^r)} l_i^r (x,u,t)_i$$

Check the stopping criterion, if it is satisfied stop, else go Step 4.

The presumption of the following proposition will be the basis for the validation of the globality of DISOPE algorithm which is used as a local search for CAM.

Proposition 3: presume that the function g has a finite number of stationary vectors. Then the CAM algorithm terminates after a finite number of iterations at a global minimizer of g [16].

A main feature of the combination of CAM and DISOPE algorithm is that the combination converges to global optimum based on the Proposition 3. Moreover, the ability of CAM is that it can be combined with a local search and will converge to a global solution. But if the output of the local search was global in the first place, the output of the combination with CAM will be the same value and that validates the globality. Therefore, it is inferred that DISOPE algorithm converges to global solution.

The examples of next section were solved by the above algorithm to show that the globality of the combination of CAM and DISOPE algorithm is valid.

4. Numerical Examples

Example 1: First minimization OCP test.

$$J = \int_0^{0.78} (x_1^2(t) + x_2^2(t) + 0.1u^2(t))dt$$

$$\text{s.t } \dot{x}_1 = -(x_1(t) + 0.25) + (x_2(t) + 0.5)\exp\left(\frac{25x_1(t)}{x_1(t)+2}\right) - (1 +$$

$$u(t))(x_1(t) + 0.25),$$

$$\dot{x}_2 = 0.5 - x_2(t) - (x_2(t) + 0.5)\exp\left(\frac{25x_1(t)}{x_1(t)+2}\right)$$

$$x(0) = [0.05\ 0]^{\mathrm{T}}$$

For this problem, the interval $[0, 0.78]$ was partitioned into 80 equal subintervals. The combination is used to solve this example via UVCT (see Figures 1 and 2). Finally the algorithm arrives at the following solution.

$$x_1{}^* = (0.0323, 0.0161, \dots, 0.323, 0.0010)_{1\times 79}$$

$$x_2{}^* = (0.0323, 0.0161, \dots, 0.323, 0.0010)_{1\times 79}$$

$$u^* = (-9.4278, -4.9996, \dots, -9.4278, -0.3170)_{1\times 79}$$

$$J^* = 0.1111$$

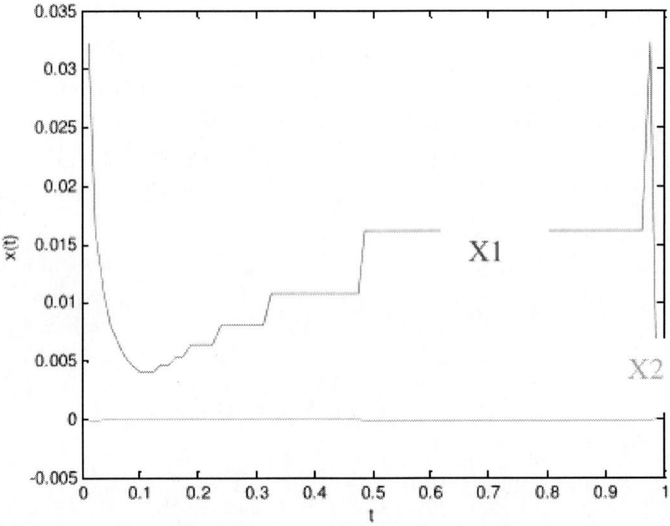

Figure 1. Optimum state vectors of Example 1 for $n = 80$.

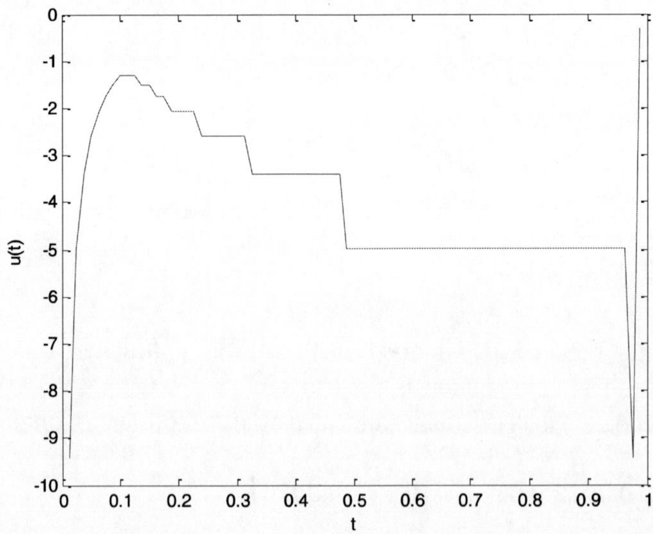

Figure 2. Optimum control vector of Example 1 for $n = 80$.

Example 2: Consider the following OCP

Minimize $J = \int_0^1 u^2(t)dt$

s.t $\dot{x}(t) = x^2(t) + u(t)$

$x(0) = 0, x(1) = 0.5$

where $n = 102$ and step size $\mathfrak{h} = 0.0099$. For this problem, the interval $[0, 1]$ was partitioned into 102 equal subintervals. The combination of CAM and DISOPE algorithm is used to solve this example via UVCT. The global solution of the discretized OCP is $J^* = 0.000055$ at

$$x = (0.09, 0.090045164478881, 0.090081894249267, \ldots, 0.082453491544067)_{1\times101}$$

with two iterations of the CAM (see Figures 3 and 4).

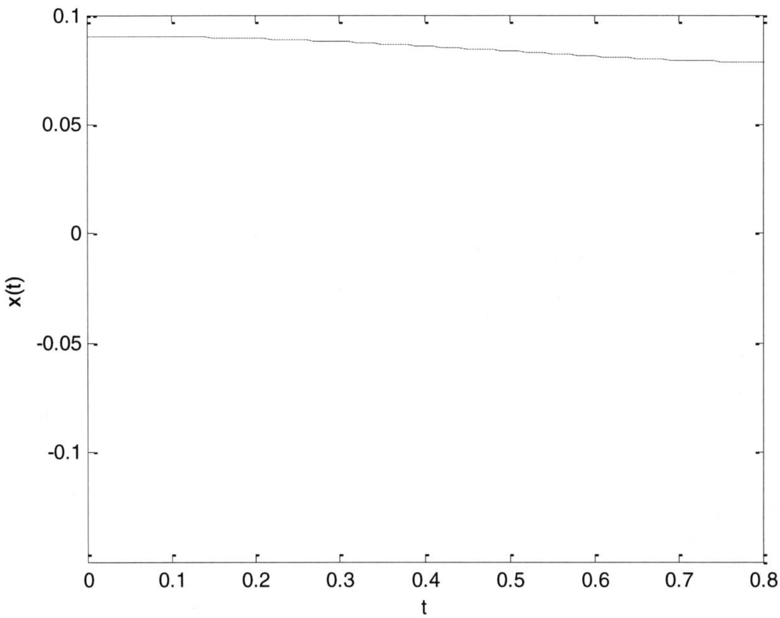

Figure 3. Optimum state vectors of Example 2 for $n = 102$.

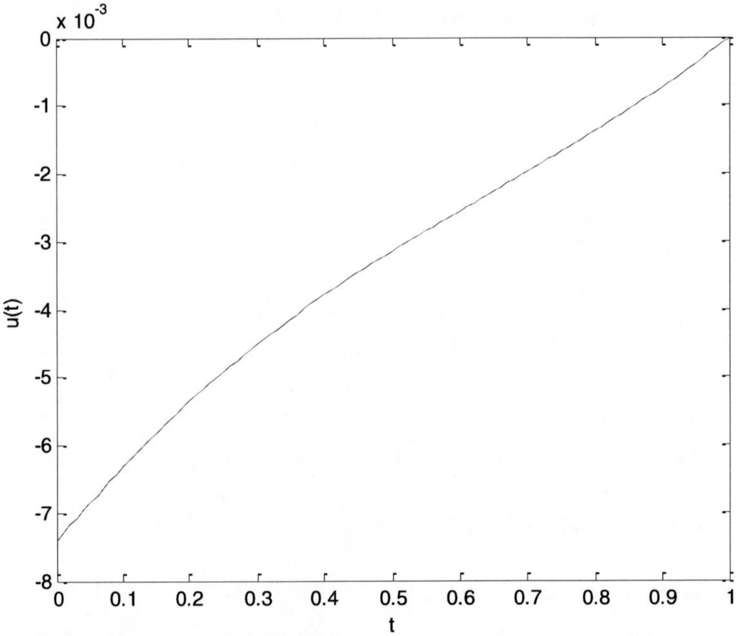

Figure 4. Optimum control vectors of Example 2 for $n = 102$.

Example 3: Consider the following OCP

Minimize $J = \int_0^1 u^2(t)dt$

s.t $\dot{x}(t) = \frac{1}{2}x^2(t)\sin(x(t)) + u(t)$

$x(0) = 0, x(1) = 0.5$

where $n = 102$ and step size $\mathfrak{h} = 0.0099$. For this problem, the interval $[0, 1]$ was partitioned into 102 equal subintervals. The Combination is used to solve this example via UVCT. The global solution of the discretized OCP is $J^* = 0.0000005$ at

$x =$
$(0.09, 0.090004070867621, 0.090008133407132, \ldots, 0.091426903511376)_{1\times 101}$

with two iterations of CAM (see Figures 5and 6).

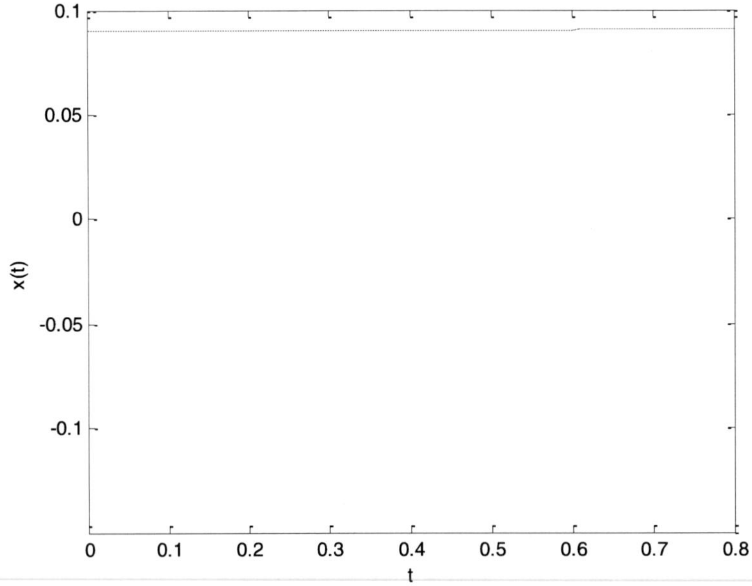

Figure 5. Optimum state vectors of Example 3 for $n = 102$.

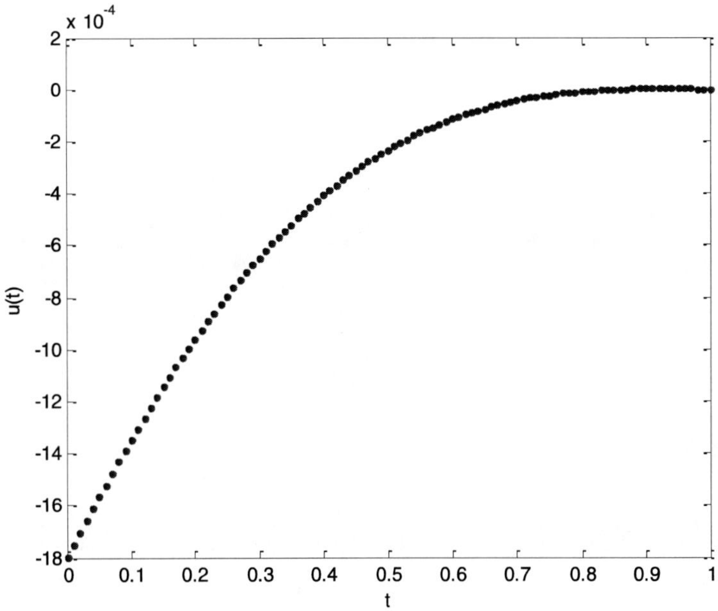

Figure 6. Optimum control vectors of Example 3 for $n = 102$.

Example 4: Consider the following OCP

Minimize $J = \int_0^1 (x^3(t) + u^2(t))dt$

s.t $\dot{x}(t) = u(t)$

$x(0) = 0, x(1) = 0.5$

where $n = 102$ and step size $\mathfrak{h} = 0.0099$. For this problem, the interval $[0, 1]$ was partitioned into 102 equal subintervals. The combination is used to solve this example via UVCT. The global solution of the discretized OCP is $J^* = 25.04115$ at

$$x = (4, 3.922888015459106, 3.848168398143530, \ldots, 1.467530434143723)_{1\times 101}$$

with two iterations of the CAM (see Figures 7 and 8).

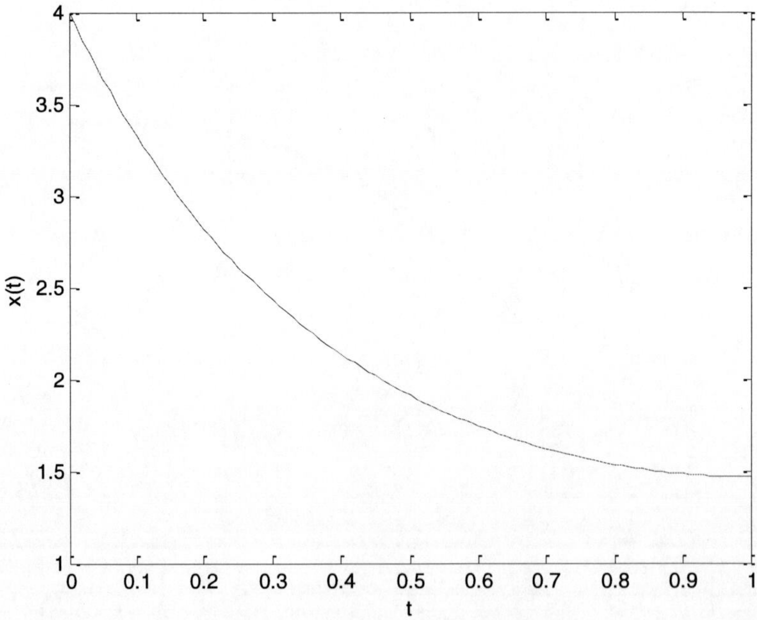

Figure 7. Optimum state vectors of Example 4 for $n = 102$.

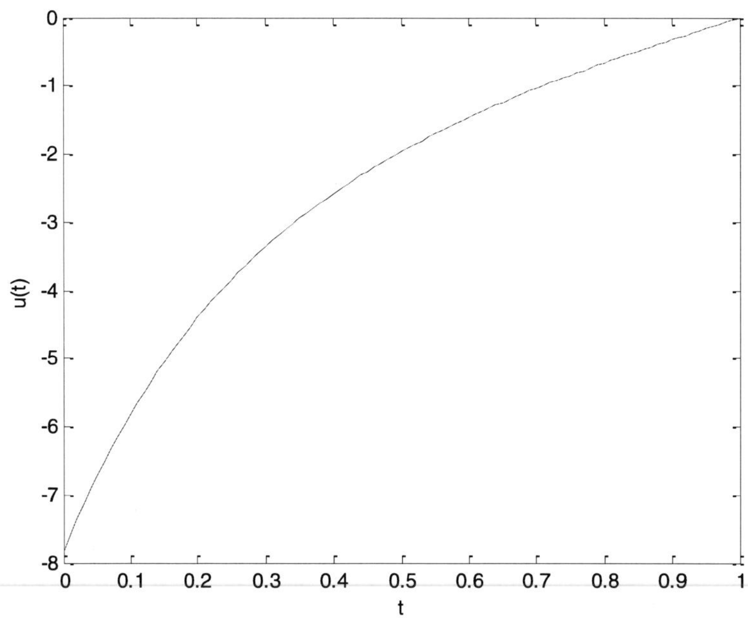

Figure 8. Optimum control vectors of Example 4 for $n = 102$.

CONCLUSION AND SUGGESTION

Since the CAM has this property to be used with a local search as a combination, it converges to global solution. Therefore, the output of the combination of CAM and DISOPE algorithm via UVCT is global. Furthermore, the numerical findings of this study and proposition 3 verify this globality. Particularly, when CAM is used as a global search together with a local search, if the output of the local search was global in the first place, then the output of the combination would also be the same value and hence global.

For further research in the application of CAM on functionals to decrease the time taken for convergence the following suggestions are presented,

i. In order to decrease the number of vectors combinations in UVCT, it is better to substitute the vectors involved for the performance index. Specially, in some examples it is more appropriate to use only the state vectors (or control vectors) combinations to converge to the global optimum with reference to the constraints and the objective functionals.

ii. Depending on the problems' forms, it is advisable to find the suitable number of partitions experimentally in discretized OCP which is a main element for converging to global solution rapidly.

iii. In some examples, choosing the proper stopping criterion shall help CAM to converge to best solution in OCP.

REFERENCES

[1] Singh M.G., and Titli A.B., *Systems: Decomposition, Optimization and Control*, Oxford: Pergamon Press, (1978).

[2] Roberts P.D., *An algorithm for optimal control of nonlinear systems with Model-Reality Differences*, 12th IFAC World Congress on Automatic Control, (1993).

[3] Becerra V.M., *Development and Applications of Novel Optimal Control Algorithms*, Dept. of Electrical, Electronics and information Engineering, City University, UK: PhD thesis, (1994).

[4] Bagirov A.M., and Rubinov A.M., *Global minimization of increasing positively homogeneous functions over the unit Simplex*, Annals of Operations Research, Netherlands: Kluwer Academic Publications, (2000).

[5] Andramonov M., Rubinov A., and Glover B., *Cutting Angle Methods in Global Optimization*, Applied Mathematics Letters 12, Elsevier Science Ltd, (1999).

[6] Roberts P.D., An Algorithm for Steady State System Optimization and Parameter Estimation, *International Journal of System science*, 10, 719-734, (1979).

[7] Roberts P.D., and Williams T.W.C, On an Algorithm for Combined System Optimization and Parameter Estimation, *Automatica*, 17, 199-209, (1981).

[8] Ellis J.E., and Roberts P.D., Simple Model for Integrated Optimization and Parameter estimation, *International Journal of Systems Science*, 12, 442-472, (1981).

[9] Stevenson I.A., Brdys M., and Roberts P.D., *Integrated System Optimization and Parameter Estimation for Travelling Load Furnace Control*, In: Barker, H.A. and Young, P.C. eds, Identification and System Parameter Estimation, Oxford: Pergamon Press, (1985).

[10] Bryds M., and Roberts P. D., Convergence and Optimality of Modified Two-Step Algorithm for Integrated System Optimization and Parameter Estimation, *International Journal of System Science*, 18, 1305-1322, (1987).

[11] Brdys M., Ellis J. E., and Roberts P. D., Augmented Integrated System Optimization and Parameter Estimation Technique: Derivation, Optimality and Convergence, *IEE Proceedings* Part 3, 14(3), 201-209, (1987).

[12] Roberts P. D., *Unit Memory Repetitive Process Aspects of Iterative Optimal Control Algorithms.* Proceedings of 33rd Conference on Decision and Control, December 14-16, IEEE, 2, 1394-1399, Lake Buena Vista, Florida, (1994).

[13] Becerra V. M., and Roberts P. D., Dynamic Integrated System Optimization and Parameter for Discrete Time Optimal Control of Nonlinear Systems, *Int. J. Control*, 63(2), 257-281, (1996).

[14] Becerra V. M., and Roberts P. D., Application of Novel Optimal Control Algorithm to a Benchmark Fed-Batch Fermentation Process. *Trans. Inst. MC*, 20(1), 11-18, (1998).

[15] Roberts P. D., and Becerra V. M., *Optimal Control of Nonlinear Systems Represented by Differential Algebraic Equations*, Proceedings of the American Control Conference, Chicago, Illinois, 762-763, (2000).

[16] Bagirov A. M., and Rubinov A. M., Cutting Angle Method and a Local Search. *Journal of Global Optimization*, Netherlands: Kluwer Academic Publishers, (2003).

[17] Seyedi S., Rohanin A., and Mohd Ismail A. A., *Capability of the Function Optimization Algorithms for Solving Functionals*, ICMAE'10, August 2010, Kuala Lumpur, Malaysia.

[18] Rubinov A. M., *Abstract Convexity and Global Optimization*, Published by Kluwer Academic Publishers, (2000).

In: A Note on Inheritance …
Editor: Iraj Sadegh Amiri

ISBN: 978-1-63485-784-0

Chapter 5

THE CUTTING ANGLE METHOD FOR SOLVING OF VARIATIONAL PROBLEMS

***Seyedalireza Seyedi*[1,*], *Iraj Sadegh Amiri*[2,†], *Sara Chaghervand*[3,‡], *Rohanin Ahmad*[4,§] and *Mohd Ismail abd Aziz*[4,‖]**

[1]Department of Mathematics, Hamedan Branch, Islamic Azad University, Hamedan, Iran
[2]Photonics Research Centre, University of Malaya, Kuala Lumpur, Malaysia
[3]Department of Electrical Engineering, Borujerd Branch, Islamic Azad University, Borujerd, Iran
[4]Department of Mathematics, Faculty Science, UniversitiTeknologi Malaysia, Skudai, Johor, Malaysia

ABSTRACT

Calculus of variations which is a field of mathematics deals with functionals. This study puts forward an application of the cutting angle method on a class of problems referred to as the variational problems.

* E-mail: seyedalireza.seyedi@gmail.com.
† E-mail: amiri@um.edu.my.
‡ E-mail: engsara1@yahoo.com.
§ E-mail: rohanin@utm.my.
‖ E-mail: mismail@mel.fs.utm.my.

The solution approach is based on a recent technique which is called the unit vectors combinations. At the end of the study the numerical findings will be presented.

Keywords: variational problem, optimization, cutting angle method

1. Introduction

In order to analyse and solve Variational problems (VPs), some main devices like calculus of variations and functional analysis are required. Generally, these are the main tools in the study of local optimization of VPs. The gradient and different types of generalized derivatives are considered really useful in the study of local extrema. As it is not feasible to resolve many global optimization issues just by local approximation, it is a necessity to develop special global devices for solving these types of problems.

There are no specific algorithms for solving any types of optimization problems. However, many optimization algorithms have been developed to solve a few varieties of which. Additionally, considering nonlinear side of most real life problems, numerous optimization algorithms have been generated to resolve nonlinear optimization issues. These problems are divided into two types of global and local optimization algorithms. The former has the ability to find global optimal solutions if sufficient time is offered whereas the latter, local optimization algorithms, does not. Local optimization algorithms include gradient descent, Newton's method, quasi Newton's method, conjugate gradient methods, etc. [1, 2]. Some primary points refer that these types are merged to create a local optimum. Such a local optimum is globally optimal as long as the objective function is quasi-convex and viable region is convex, which is a scarcity for them to occur practically [2, 3, 4, 5, 6, 7, 8]. Global optimization algorithms incorporate mechanisms which are able to avoid local optimization. This paper focuses on a type of global optimization called Cutting Angle Method (CAM) which is a generalization of the cutting plane method from convex minimization driven from findings on abstract convexity [9]. In [10], this method was introduced and later on [9] it was expanded to minimize the increasing convex-along-rays functions as a special case. Despite that it can be applied to an inclusive class of non-convex global optimization problems in which the functions are involved. It possesses suitable generalized affine minorants [10].

This study is going to introduce an application of CAM on a class of problems referred to as the variational problems (VPs) which comprise of two steps:

Step1: Discretize the VP, and

Step2: Use the CAM to solve the Discretized problem through new technique called Unit Vectors Combinations Technique (UVCT).

Furthermore, convergence of the solutions of the Discretized VPs to extremal via CAM is based on the convergence of the CAM to global solution by [10].

2. An Application of the Cutting Angle Method on Variational Problems

Through discretization techniques as well as the Inheritance and Generalizability of function optimization algorithms for solving functionals which were proposed by [11] the next procedure shall be examined to introduce an application of CAM on functionals.

Procedure 1

III. (Discretization): The original problem in this research is in the form of:

$$\begin{aligned} &\text{Minimize } J = \int_{t_0}^{t_1} F(y, y', t)dt \\ &s.t\ \dot{y} = g(y, y', t) \\ &y(t_0) = y_0, y(t_1) = y_1, \end{aligned} \tag{1}$$

where $F: \mathcal{A} \times \mathcal{B} \times [t_0, t_1] \to \mathbb{R}^n$ is a continuousfunction, g is a continuous nonlinear time varying function, $g: \mathcal{A} \times \mathcal{B} \times [t_0, t_1] \to \mathbb{R}^n, t \in [t_0, t_1] \subseteq \mathbb{R}, y(t) \in \mathcal{A}, y'(t) \in \mathcal{B}$, and $\mathcal{A}, \mathcal{B}$ are compact subsets of $\mathbb{R}^n$ which must be chosen. So, the system reaches from initial state $y(t_0)$ to final state $y(t_1)$.

Firstly, discretize the original problem (1) to the following equivalent problem and partition intervals $[t_0, t_1]$ to$n$ parts and also $t \in [t_0, t_1]$.

$$\text{Minimize } J = \mathfrak{h}\sum_{i=1}^{n-1} F_i\,(y, y', t) \tag{2}$$

$$s.t\ y = \mathfrak{h} * g(y, y', t)$$
$$y(t_0) = y_0, y(t_1) = y_1, \tag{3}$$

where $\mathfrak{h}$ is a step size of the discretization, $\mathfrak{h} = t_1 - t_0/n$, $y = \{y_j\}, y' = \{y'_k\}$, and $k = 1,2,\dots,n$.

IV. Secondly, apply CAM for solving the discretized VP (2) via UVCT as follows:

In Step 0 of CAM, initialize the state vector y_1 by unit vectors

$$e^1 = (1,0,0,\dots,0,0)_{1\times n-1}, e^2 = (0,1,0,\dots,0,0)_{1\times n-1}, \dots,$$

$$e^{n-2} = (0,0,0,\dots,1,0)_{1\times n-1}, e^{n-1} = (0,0,0,\dots,0,1)_{1\times n-1}.$$

Next, calculate the corresponding values of $y_2, \dots, y_m; y'_1, y'_2, \dots, y'_m$ via (3) and calculate the corresponding value of J_1 via (2), where $y' = \frac{y(t_k+\mathfrak{h})-y(t_k)}{\mathfrak{h}}$, $m = 1, 2, \dots, n-1$. Then, construct basis vectors

$$L^p = \left(\frac{J(e^p, y', t)}{e^p}\right), p = 1, \dots, n-1.$$

And apply CAM to find y_1^* based on support vectors $L^j = \left(\frac{J(y^*, y', t)}{y^*}\right)$, $j = the\ number\ of\ iterations\ of\ CAM$. Then calculate the corresponding values of other components via (3) and J_1^*.

Next, continue this operation for the state vectors $y_2, y_3, \dots, y_m$ to find $y_2^*, y_3^*, \dots, y_m^*$ and corresponding optimal values $J_2^*, J_3^*, \dots, J_m^*$.

The UVCT initializes 2, 3 and $m-1$ sate vectors combinations respectively by unit vectors and use CAM to find corresponding optimal values of them in parallel as follows.

CAM can be applied for one vector or combination of two, three or more states vectors:

$y_1, y_2; y_1, y_3; \ldots; y_1, y_m; y_2, y_1; y_2, y_3; \ldots;$
$y_2, y_m, \ldots y_1, y_2, y_3; \ldots; y_1, y_2, y_m;$
$\ldots; y_m, y_{m-1}, \ldots, y_2$

and find the corresponding optimal solutions

$J_{12}^*, J_{13}^*, \ldots, J_{1m}^*; J_{21}^*, J_{23}^*, \ldots, J_{2m}^*;$
$\ldots J_{123}^*, \ldots, J_{12m}^*; \ldots; J_{mm-1\ldots 2}^*$

Regarding the minimization of the discretized VP by CAM, the set

$$\mathcal{J}^* = \{J_1^*; J_2^*, J_3^*, \ldots, J_m^*; J_{12}^*, J_{13}^*, \ldots, J_{1m}^*; J_{21}^*, J_{23}^*, \ldots, J_{2m}^*; \ldots; J_{123}^*, \ldots, J_{12m}^*; \ldots; \ldots; J_{m\,m-1\ldots 2}^*\}.$$

will be obtained and the smallest elements of the set $\mathcal{J}^*$ will be the global optimum solution of the mentioned VP based on the global property of CAM.

3. Generalization of Cutting Angle Method for Variational Problems

Consider the following global optimization problem:

$$J(y, y', t) \to min$$

subject to $y(t), y'(t) \in \mathcal{S}^*$,

where $\mathcal{S}^* = \{t \in \mathbb{R}_+^n : \sum_i^r y_i(t) = 1, \sum_i^r y'_i(t) = 1\}$ and r is the number of coordinates of the vectors $y(t), y'(t)$.

Based on the Inheritance and Generalizability properties of function on functional, which was proposed by [12], the following expressions are represented as an extension of their function framework which was similarly presented by [9, 13].

Let $G: \mathcal{S}^* \to \mathbb{R}$ be a positive Lipschitz functional defined on the unit simplex. Then it can be extended to a finite IPH functional $J(y, y', t)$ on the cone $\mathbb{R}_+^n$, which would coincide with $G(y, y', t) + c$ on $\mathcal{S}^*$. I.e., $J(y, y', t) = G(y, y', t) + c$ is an IPH functional on $\mathcal{S}^*$. Afterwards, for each $y(t), y'(t) \in \mathbb{R}_+^n$ define the support vectors

$$L_y = L = \left(\frac{J(y,y',t)}{y}\right) = \left(\frac{J(y,y',t)}{y_1}, \frac{J(y,y',t)}{y_2}, \dots, \frac{J(y,y',t)}{y_n}\right).$$

The n vectors $e^m = (0, \dots, 0,1,0, \dots, 0)$ with 1 in the mth position shall be used, and the corresponding support vectors $L^m = \left(\frac{J(e^m,y',t)}{e^m}\right)$, $m = 1, \dots, n$ shall be called basis vectors. A set of $k \geq n$ supports vectors (and hence k known values of the functional $J(y,y',t)$ at k distinct vectors) $\mathcal{K}_y = \{L^k\}_{j=1}^{k}$. Let also the first n support vectors be the basis vectors. This choice of support vectors guarantees that the algorithm will locate all local (and hence global) minimizers of the auxiliary functionals

$$H_k(y,y',t) = \max_{j \geq k} \min_{i=1,\dots,n} L_i^j y_i$$

in the interior of the unit simplex. It always underestimates the value of $J(y,y',t)$: $H_k(y,y',t) \leq J(y,y',t)$. Hence, $b_k = \min_{y \in S^*} H_k(y,y',t) \leq \min_{y \in S^*} J(y,y',t)$. On the other hand, the sequence of its minima, $\{b_k\}_{k=n}^{\infty}$ is increasing, and convergence to the global minimum of $J(y,y',t)$is based on its function framework. Based on the CAM formulated by [9] on function, the following algorithms are presented for VPs.

The CAM for VP

Algorithm 1

Step (0). (Initialization)

(d) Take points e^m, $m = 1, \dots, n-1$, and construct basic vectors $L^m = \left(\frac{J(e^m,y',t)}{e^m}\right)$, $m = 1, \dots, n-1$.

(e) Define the functional $H_n(x) = \max_{j \leq n} \min_{i=1,\dots,n} L_i^j y_i = \max_{j \leq n} L_j^j y_j$.

(f) Set $k = n - 1$.

Step 1. Find $y^* = \arg\left[\min_{y \in S^*} H_k(y,y',t)\right]$.

Step 2. Set $k = k + 1$and $y^k = y^*$.

Step 3. Compute $L^j = \left(\frac{J(y^k,y',t)}{y^k}\right)$. Define the function

$H_k(y,y',t) = \max_{j \le k} \min_{i=1,\dots,n} L_i^j y_i =$
max{H_(k − 1) (y, y^', t), min┬(i = 1, …, n) 〖L_i^k y_i〗}.

Go to Step 2.

Minimization of the auxiliary functional, the step 1 of the algorithms 1 and 2, is considered as one of the most time-consuming steps. To resolve this problem, the following theorem is presented.

Theorem 1: Let $y^*(t) > 0$ be a local minimizer of $H_k(y,y',t)$ over the relative interior of S^*,$riS^* = \{y(t) \in S^*, y(t) > 0\}$. Then there exists a subset $\mathcal{L} = \{L^{j_1}, L^{j_2}, \dots, L^{j_n}\}$ of the set $\mathcal{K}$, such that

(1) $y = \{d/L_1^{j_1}, d/L_2^{j_2}, \dots, d/L_n^{j_n}\}$ with $d = (\sum_i 1/L^{j_i})^{-1}$
(2) $\max_{j \le k} \min_{i=1,\dots,n-1} L_i^j / L_i^{j_i} = 1.$
(3) Either$\forall i: j_i = i$, or$\exists m: k_m > n, L_i^{j_m} > L_i^{j_i}, \forall i \neq m.$

Proof: Completely similar to the proof on function by [14].

The value of the auxiliary functional at $y$is $H_k(y,y',t) = d$.

The following version of CAM for VPs works based on the above theorem, by analysing all possible combinations of n support vectors.

Consider a set of k support vectors $\mathcal{K} = \{L^j\}_{j=1}^k$, $L^j \in \mathbb{R}_+^n$. Let $\mathcal{I}$ denote $\{1,2,\dots,n\}$. From Theorem 7.1, the local minima of the auxiliary functional $H_k(y,y',t)$are combinations of n support vectors $\mathcal{L} = \{L^{j_1}, L^{j_2}, \dots, L^{j_n}\}$ that satisfy the following conditions:

(III)$\forall i, s \in \mathcal{I}, i \neq s: L_i^{j_i} < L_i^{j_s}$;
(IV) $\forall v \in \mathcal{K} \backslash \mathcal{L}, \exists i \in \mathcal{I}: L_i^{j_i} \ge v_i.$

The subset $\mathcal{L}$, satisfying conditions (I) and (II) above, is called a valid combination of support vectors.

Let $\mathcal{W}^k$ denote the set of all valid combinations $\mathcal{L}$ of k support vectors, satisfying conditions (I) and (II):

$$\mathcal{W}^k = \{\mathcal{L} = \{L^{j_1}, L^{j_2}, \dots, L^{j_n}\}, L^{j_i} \in \mathcal{K}: (1) \forall i, s \in \mathcal{I}, i \neq s: \mathcal{L}_{ii} < \mathcal{L}_{si}$$

and

(II) $\forall v \in \mathcal{K} \backslash \mathcal{L}\, \exists i \in \mathcal{I}: \mathcal{L}_{ii} \geq v_i\}$

The problem of finding local minima of $H_k(y, y', t)$ is transmuted into the problem of listing the elements of $\mathcal{W}^k$. A simplistic approach is then:

Step 1. Construct all combinations $\mathcal{L}$ satisfying (I)
Step 2. Check the obtained combinations against (II).

Then improve on this by taking this fact into account that all elements of $\mathcal{W}^k$ not involvingL^j (i.e., minima of $H_{k-1}(y, y', t)$) do not need to be recomputed; hence, the algorithm takes the following form:

Algorithm 2

Step (0). (Initialization)

(a) Take points e^m, $m = 1, \dots, n-1$, and construct basic vectors $L^m = \left(\frac{J(e^m, y', t)}{e^m}\right)$, $m = 1, \dots, n$.
(b) Define the function $H_n(y, y', t) = \max_{j \leq n} \min_{i=1,\dots,n} L_i^j y_i = \max_{j \leq n} L_j^j y_j$.
(c) Set $k = n$. Set $\mathcal{W}^k = \{\{L^1, L^2, \dots, L^{n-1}\}\}$.
(d) Calculate $d = (\sum_{i=1,\dots,n-1} 1/L_i^i)^{-1}$.

Step 1

(a) Retrieve all valid combinations $\mathcal{L}$ (i.e., $\mathcal{W}^k$).
(b) Select $\mathcal{L} \in \mathcal{W}^k$.

Step 2

(d) $k = k + 1$.
(e) Take $y^* = d/\text{diag}(\mathcal{L})$ and evaluate $J(y^*, y', t)$.
(f) Compute $L^k = (J(y^*, y', t)/y^*)$. Define the function

$$H_k(y, y', t) = \max_{j \leq k} \min_{i=1,\dots,n} L_i^j y_i =$$
$$\max\left\{H_{k-1}(y), \min_{i=1,\dots,n} L_i^k y_i\right\}$$

Step 3

(a) Check $\mathcal{W}^{k-1}$ against (II) and remove those that fail (II).
(b) Move the remaining combinations into $\mathcal{W}^{k}$.

Step 4

(a) Construct all combinations $\mathcal{L}$ that involve L^{k} and satisfy (I).
(b) Calculate $d = (\sum_{i=1,\ldots n} 1/L_i^{k_i})^{-1}$ for each such combination.
(c) Add these combinations to $\mathcal{W}^{k}$.
(d) Go to Step 1.

Remark 1: In Algorithm 2, if the diagonal of $\mathcal{L}$was dominated byL^{k}, this combination would not be moved into$\mathcal{W}^{k+1}$.

Remark 2: In Algorithm 2, in any iteration, the algorithm computes and checks the value of $J(y^*, y', t) - d$ with the value of assumed stopping criteria. While it satisfies then the algorithm stopped.

It is obvious that the set $\mathcal{J}^*$ is countably finite. Furthermore, with regard to the following proposition, $\mathcal{J}^*$ is not empty:

Theorem 2: Suppose that the function f is continuous on the closed interval $[a, d]$. If $\mathfrak{m}$ and $\mathfrak{n}$ are, respectively, the absolute minimum and absolute maximum function values of f on $[a, d]$ so that [15],

$$\mathfrak{m} \leq f(x) \leq \mathfrak{n} \text{for} a \leq x \leq d$$

then

$$\mathfrak{m}(d - a) \leq \int_a^d f(x)\, dx \leq \mathfrak{n}(d - a).$$

Theorem 3 (The Mean-Value Theorem for Integrals): If f is continuous on the closed interval $[a, d]$, there exists a number c in $[a, d]$ such that [15],

$$\int_a^d f(x)\, dx = f(c)(d - a).$$

Proposition 1: The set of all optimal solutions $\mathcal{J}^*$ by CAM for discretized OCP is not empty.

Proof: the original OCP can be shown as follows

$$J(y, y', t) = \int_a^d F(y, y', t)\, dt =$$

$$\int_a^d F_1(y,y',t)\,dt + \int_a^d F_2(y,y',t)\,dt + \cdots + \int_a^d F_n(y,y',t)\,dt$$

where

$$F(y,y',t) = F_1(y,y',t) + F_2(y,y',t) + \cdots + F_n(y,y',t).$$

Based on Theorems 2 and 3 the following assertions shall be presented

$$\mathfrak{m}_1(d-a) \le \int_a^d F_1(y,y',t)\,dt = F_1(c_1)(d-a)$$
$$\mathfrak{m}_2(d-a) \le \int_a^d F_2(y,y',t)\,dt = F_2(c_2)(d-a)$$

$$\mathfrak{m}_n(d-a) \le \int_a^d F_n(y,y',t)\,dt = F_n(c_n)(d-a) \tag{4}$$

Next, (4) implies that

$$\begin{aligned}
&\mathfrak{m}_1(d-a) + \mathfrak{m}_2(d-a) + \cdots + \mathfrak{m}_n(d-a)\\
&\le \int_a^d F_1(y,y',t)\,dt + \int_a^d F_2(y,y',t)\,dt + \cdots + \int_a^d F_n(y,y',t)\,dt\\
&= F_1(c_1)(d-a) + F_2(c_2)(d-a) + \cdots + F_n(c_n)(d-a)\\
&\Rightarrow (\mathfrak{m}_1 + \mathfrak{m}_2 + \cdots + \mathfrak{m}_n)(d-a) \le J(y,y',t) = c(d-a)\\
&\Rightarrow \mathfrak{m}(d-a) \le J(y,y',t) = c(d-a)
\end{aligned}$$

where

$\mathfrak{m} = \mathfrak{m}_1 + \mathfrak{m}_2 + \cdots + \mathfrak{m}_n$ and $c = F_1(c_1) + F_2(c_2) + \cdots + F_n(c_n)$

Thus

$$\mathfrak{m} \le c$$

Then there exists a $q \in \mathbb{R}$ such that $q = \mathfrak{m}$ or $q = c$or, $\mathfrak{m} < q < c$. Then implies that

$$J_q \le J(y,y',t) \Rightarrow J_q \in \mathfrak{J}.$$

where

$q\,(d-a) \le J_q = J(y_q, y'_q, t)$ or $J_q = J\left(y_q, y'_q, t\right) = q(d-a)$.

Thus $J_{q_i} \in \mathcal{J}^*$ and the set $\mathcal{J}^*$ is not empty.

Remark 3: Moreover, this proposition represents that the global optimum of functional J belongs to the set $\mathcal{J}^*$ constantly.

With regard to the Inheritance and Generalizability of function optimization algorithm for solving functional [11], Proposition 1and Remark 3 display that the globality of CAM to solve functions can be broadened in order to solve functional. This implies that the output of CAM for solving functional via UVCT is global. The following examples shall illustrate validity of the theme.

4. Numerical Results

Example 1: The first minimization test functional is

$$J = \int_0^1 (y_1^2(t) + y_2^2(t))dt$$

$$s.t\ \dot{y}_1 = y_2(t), \dot{y}_2 = 10y_1^3(t) + y_3(t)$$

$$y_1(0) = 0, y_1(1) = 0.1,$$

$$y_2(0) = 0, y_2(1) = 0.3.$$

For this problem, the interval $[0, 1]$ was partitioned into 10 equal subintervals. So, for $n = 10{:}\ t_1 = 0, t_2 = 0.1, \ldots, t_{10} = 1$. Then CAM was applied to three iterations to minimize the mentioned VP and Criterion= 0.001. For this example, the CAM is used to find $y_1{}^*$, $y_2{}^*$ and $y_3{}^*$ based on UVCT (see Figure 1). The global solution of the discretized VP is 0.0218 by three iterations of CAM. The following tables illustrate the convergence of CAM to the global solution with low number of iterations.

Example 2: The second minimization test functional is

$$J = \int_0^1 (y_2{}^2(t) + y_1(t))dt$$

$$s.t\ \dot{y}_2(t) = y_1(t),$$

$$y_1(0) = 0, y_1(1) = 0.5,$$

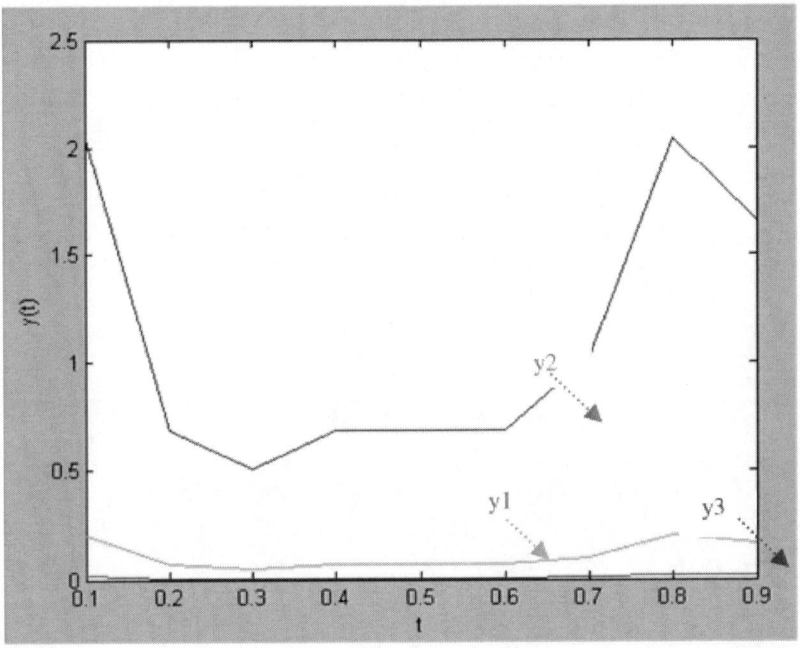

Figure 1. Optimum vectors of Example 1 for $n = 10$.

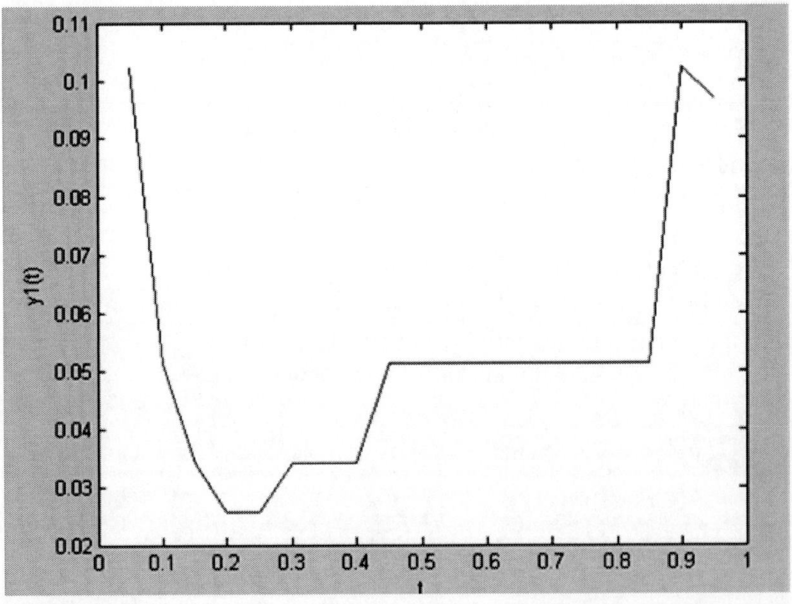

Figure 2. (Continued).

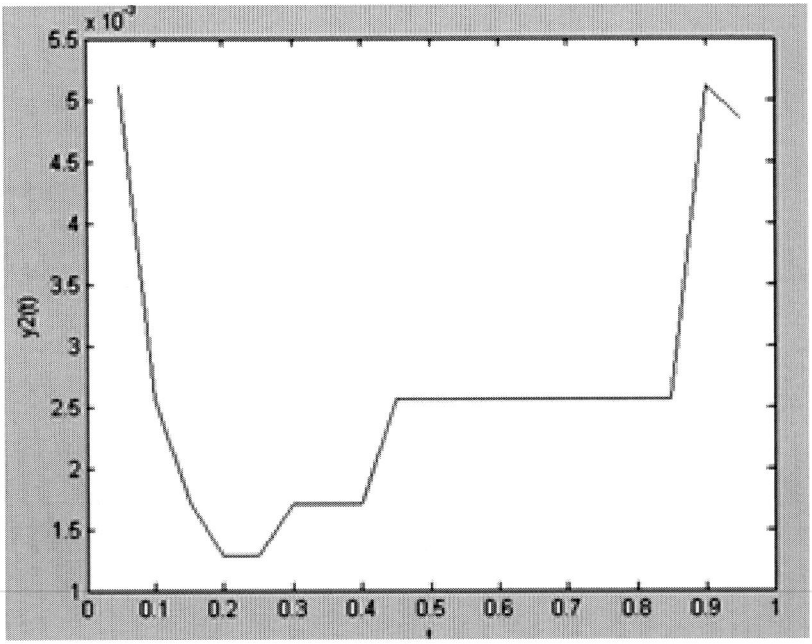

Figure 2. Optimum vectors of Example 2 for $n = 20$.

For this problem, the interval $[0, 1]$ was partitioned into 20 equal subintervals. So, for $n = 20$: $t_1 = 0, t_2 = 0.05, \ldots, t_{20} = 1$. Then, CAM is applied to three iterations to minimize the mentioned VP with stopping criterion $= 0.001$. Thus, the global minimizer is $J^* = 0.0056$ (see Figures 2).

Example 3: The third minimization test functional is

$$J = \int_0^1 (y_2{}^2(t))dt$$

$$s.t\ \dot{y_1}(t) = y_1{}^2(t) + y_2(t),$$

$$y_1(0) = 0, y_1(1) = 0.5,$$

For this problem, the interval $[0, 1]$ was partitioned into 10 equal subintervals. So, for $n = 10$: $t_1 = 0, t_2 = 0.1, \ldots, t_{10} = 1$. Then CAM used two iterations to minimize the mentioned VP and Criterion $= 0.01$.

For this example, the CAM is used to find $y_1{}^*$ and $y_2{}^*$ based on UVCT.

$$y_1{}^* =$$

$(0.0002, 0.0001, 0.0000, 0.0001, 0.0001, 0.0001, 0.0001, 0.0002, 0.0002)$

$y_2^* =$
$(0.1998, 0.0666, 0.0499, 0.0666, 0.0666, 0.0666, 0.0999, 0.1998, 0.1842)$

The global solution of the discretized VP is 0.0214 by two iterations of CAM.

Conclusion

This paper introduced and examined a direct CAM to solve variational problems via UVCT. The focal finding of this research indicates that CAM can converge to the best solution in the lowest number of iterations.

References

[1] Baldi P., Gradient Decent Learning Algorithm Overview: A General Dynamical Systems Perspective, *IEEE Transactions on Neural Networks*, 6(1), 182-195, (1995).
[2] Shang Y., *Global Search Method for Solving Nonlinear Optimization Problems*, University of Illinois at Urbana-Champaign: PhD Thesis, (1997).
[3] Stoer J., and Witzgall C., *Convexity and Optimization in Finite Dimensions I*, Berlin: Springer-Verlag, (1970).
[4] Koo, *Elements of Optimization: with Applications in Economics and Business*, New York: Springer-Verlag, (1977).
[5] Cesari L., *Optimization: Theory and Applications: Problems with Ordinary Differential Equations*. New York: Springer-Verlag, (1983).
[6] Bunday B., *Basic Optimization Methods*, London: Edwards Arnold, (1984).
[7] Rao S., *Optimization: Theory and Application*, New Delhi: Wiley Eastern Limited, (1984).
[8] Beal E., *Introduction to Optimization*, Published by John Wiley & Sons: New York, (1988).
[9] . Rubinov A. M, *Abstract Convexity and Global Optimization*, Published by Kluwer Academic Publishers, (2000).

[10] Andramonov M., Rubinov A., and Glover B., *Cutting Angle Methods in Global Optimization*, Applied Mathematics Letters12, Elsevier Science Ltd, (1999).

[11] Seyedi S., Rohanin A., and Mohd Ismail A. A., *Capability of the Function Optimization Algorithms for Solving Functionals*, ICMAE'10, August 2010, Kuala Lumpur, Malaysia.

[12] Seyedi S., Rohanin A., and Mohd Ismail A. A., *Inheritance of Function Properties for functional*, ICORAFSS 2009, Malaysia.

[13] Bagirov A. M., and Rubinov A. M., *Global minimization of increasing positively homogeneous functions over the unit Simplex*, Annals of Operations Research, Netherlands: Kluwer Academic Publications, (2000).

[14] Bagirov A. M., and Rubinov A. M., Cutting Angle Method and a Local Search. *Journal of Global Optimization*, Netherlands: Kluwer Academic Publishers, (2003).

[15] Lethold L., *The Calculus 7*, Published by Harper Collins College Pablishers Inc: United States of America, (1996).

In: A Note on Inheritance …
Editor: Iraj Sadegh Amiri

ISBN: 978-1-63485-784-0

Chapter 6

THE CUTTING ANGLE METHOD ON OPTIMAL CONTROL PROBLEMS

Seyedalireza Seyedi[1,*], Iraj Sadegh Amiri[2,†], Sara Chaghervand[3,‡], Rohanin Ahmad[4,§] and Mohd Ismail abd Aziz[4,#]

[1]Department of Mathematics, Hamedan Branch, Islamic Azad University,Hamedan, Iran
[2]Photonics Research Centre, University of Malaya, Kuala Lumpur, Malaysia
[3]Department of Electrical Engineering, Borujerd Branch, Islamic Azad University, Borujerd, Iran
[4]Department of Mathematics, Faculty Science, UniversitiTeknologi Malaysia, Skudai, Johor, Malaysia

ABSTRACT

A study of optimal control is a way of taking a best set of decisions to accomplish a particular objective which deals with functionals. This paper focuses on an application of the cutting angle method to a class of

* seyedalireza.seyedi@gmail.com.
† amiri@um.edu.my.
‡ engsara1@yahoo.com.
§ rohanin@utm.my.
# mismail@mel.fs.utm.my.

problems referred to as the optimal control problems. The solution approach is based on a new technique called unit vector combinations. At the end of this study the numerical findings are put forward.

Keywords: optimal control problem, optimization, cutting angle method

1. INTRODUCTION

Optimal control is a study which deals with selecting the best set of decisions to achieve a particular objective [1]. The performance of physical processes depends on a great deal of these decisions. The static process and the dynamic process are two classifications of Physical processes. The dynamic process can be further segregated into two linear and the non-linear processes. The main focus of this study is on the non-linear dynamic process.

Following modeling a linear process through optimization of the non-linear dynamic process, it is solved iteratively by applying an appropriate dynamic optimization algorithm. Despite that there are some inaccuracies in the mathematical models employed, the solution of the linear process will converge to the solution of the original non-linear problem [2, 3].

Although countless algorithms are accessible in order to solve Optimal Control Problems (OCP), there are essentially local search algorithms. To solve the problem associated with local searches, most OCP are modeled as Linear Quadratic Regulator (LQR) problems. So it is hoped that this newfound solution can be considered as the true global solution of the original problem. However, when it is done, a great deal of information which is carried by the original problem might be lost in its translation into LQR models [4].

This paper shall introduce a dynamic global search algorithm called Cutting Angle Method (CAM. This method was introduced by [5] which is a generalization of the cutting plane method from convex minimization, for solving OCP. CAM has been successfully applied on functions, whereas OCP are functionals.

According to [6], CAM can be possibly applied to a noticeably broad class of non-convex global optimization problems [5] claimed that CAM used in combination with a local search algorithm to arrive at a true global solution, has the ability to help local search algorithms to converge to global optima. Basically the approach begins with a solution of a local search algorithm. The cutting angle method is then applied and repeated until one arrives at a solution that is lower than the current solution of the local search algorithm.

While the lower solution as the new initial solution is employed, the local search algorithm is again run on the problem. As soon as a new solution is obtained from this local search, the cutting angle method is once again used and the process would not stop unless the cutting angle method fails to find a lower valued solution after a predetermined number of iterations. The newest solution of the local search algorithm would be considered as the global optimum. This research mainly concerns on proposing a technique to employ CAM for solving OCP directly.

2. An Application of the Cutting Angle Method on Optimal Control Problems

The next procedure shall analyse an application of CAM to OCP. This application is generated from two approaches which were introduced by [7] discretization techniques and the Inheritance and Generalizability of function optimization algorithms for solving functional (In Particular, OCP).

Procedure 1:

I. (Discretization): In this study, the original problem forms as:

$$\text{Minimize } J = \int_{t_0}^{t_1} E(\boldsymbol{x},\boldsymbol{u},t)dt \tag{1}$$

$$s.t\ \dot{\boldsymbol{x}} = g(\boldsymbol{x},\boldsymbol{u},t) \tag{2}$$
$$\boldsymbol{x}(t_0) = \boldsymbol{x}_0, \boldsymbol{x}(t_1) = \boldsymbol{x}_1,$$

where $E: \mathcal{A} \times \mathcal{B} \times [t_0, t_1] \to \mathbb{R}$ is a continuous function, g is a continuous nonlinear time varying function, $g: \mathcal{A} \times \mathcal{B} \times [t_0, t_1] \to \mathbb{R}^n, t \in [t_0, t_1] \subseteq \mathbb{R},$, $\boldsymbol{x}(t) \in \mathcal{A}$, $\boldsymbol{u}(t) \in \mathcal{B}$ and $\mathcal{A} \subseteq \mathbb{R}^n, \mathcal{B} \subseteq \mathbb{R}^m$ are compact subsets which must be chosen in order for the system to reach from initial state $\boldsymbol{x}(t_0)$ to final state $\boldsymbol{x}(t_1)$.

Firstly, discretize the original problem (1) to the following equivalent problem and partition intervals $[t_0, t_1]$ to n parts and also $t \in [t_0, t_1]$.

$$\text{Minimize } J = \mathfrak{h} \sum_{i=1}^{n-1} E_i\,(\boldsymbol{x},\boldsymbol{u},t) \tag{3}$$

$$s.t\ \mathrm{x} = \mathfrak{h} * g(\boldsymbol{x},\boldsymbol{u},t) \tag{4}$$

$$\boldsymbol{x}(t_0) = \boldsymbol{x}_0, , \boldsymbol{x}(t_1) = \boldsymbol{x}_1,$$

where $\mathfrak{h}$ is a step size of the discretization, $\mathfrak{h} = t_1 - t_0/n$, $\boldsymbol{x} = \{\boldsymbol{x}_j\}, \boldsymbol{u} = \{\boldsymbol{u}_k\}, j = 1,2,\ldots,m_1$ and $k = 1,2,\ldots,m_2$.

II. Secondly, to solve the discretized OCP (3) via unit vectors Combinations technique (UVCT), apply CAM. In the current study CAM will be generalized to solve discretized OCP via the UVCT as follows,

In Step 0 of CAM initialize the state vector $\boldsymbol{x}_1$ by unit vectors

$$e^1 = (1,0,0,\ldots,0,0)_{1\times n-1}, e^2 = (0,1,0,\ldots,0,0)_{1\times n-1}, \ldots,$$
$$e^{n-2} = (0,0,0,\ldots,1,0)_{1\times n-1}, e^{n-1} = (0,0,0,\ldots,0,1)_{1\times n-1}.$$

Next, calculate the corresponding values of $\boldsymbol{x}_2, \boldsymbol{x}_3, \ldots, \boldsymbol{x}_{m_1}; \boldsymbol{u}_1, \boldsymbol{u}_2, \ldots, \boldsymbol{u}_{m_2}$ via (4) and calculate the corresponding value of J_1 via (3). Then, construct basis vectors

$$L^p = \left(\frac{J(e^p, \mathrm{u}, t)}{e^p}\right), p = 1, \ldots, n-1.$$

And apply CAM to find $\boldsymbol{x}_1{}^*$ based on support vectors $L^k = \left(\frac{J(\mathrm{x}^*, \mathrm{u}, t)}{x^*}\right)$, $k =$ the number of iterations of CAM. Then calculate the corresponding values of other components via (4) and $J_1{}^*$.

Next, continue this operation for the state vectors $\boldsymbol{x}_2, \boldsymbol{x}_3, \ldots, \boldsymbol{x}_{m_1}$ to find $\boldsymbol{x}_2{}^*, \boldsymbol{x}_3{}^*, \ldots, \boldsymbol{x}_{m_1}{}^*$ and corresponding optimal values $J_2{}^*, J_3{}^*, \ldots, J_{m_1}{}^*$.

The UVCT initializes 2, 3 and $m_1 - 1$ sate vectors combinations respectively by unit vectors and use CAM to find corresponding optimal values of them in parallel as follows.

CAM can be applied for one state vector or combination of two, three or more states vectors:

$$\boldsymbol{x}_1, \boldsymbol{x}_2; \boldsymbol{x}_1, \boldsymbol{x}_3; \ldots; \boldsymbol{x}_1, \boldsymbol{x}_{m_1}; \boldsymbol{x}_2, \boldsymbol{x}_1; \boldsymbol{x}_2, \boldsymbol{x}_3; \ldots;$$
$$\boldsymbol{x}_2, \boldsymbol{x}_{m_1}, \ldots \boldsymbol{x}_1, \boldsymbol{x}_2, \boldsymbol{x}_3; \ldots; \boldsymbol{x}_1, \boldsymbol{x}_2, \boldsymbol{x}_{m_1};$$
$$\ldots; \boldsymbol{x}_{m_1}, \boldsymbol{x}_{m_1-1}, \ldots, \boldsymbol{x}_2$$

And find the corresponding optimal solutions

$$J_{12}^{*}, J_{13}^{*}, \dots, J_{1\,m_1}^{*}; J_{21}^{*}, J_{23}^{*}, \dots, J_{2\,m_1}^{*}; \\ \dots J_{123}^{*}, \dots, J_{12\,m_1}^{*}; \dots; J_{m_1\,m_1-1\dots 2}^{*}$$

respectively.

In a similar way, CAM can be used for one control vector or combination of two, three or more control vectors:

$$\boldsymbol{u}_1, \boldsymbol{u}_2, \boldsymbol{u}_3, \dots, \boldsymbol{u}_{m_1}; \quad \boldsymbol{u}_1, \boldsymbol{u}_2; \boldsymbol{u}_1, \boldsymbol{u}_3; \quad \dots; \quad \boldsymbol{u}_1, \boldsymbol{u}_{m_1}; \quad \boldsymbol{u}_2, \boldsymbol{u}_1; \boldsymbol{u}_2, \boldsymbol{u}_3; \quad \dots; \\ \boldsymbol{u}_2, \boldsymbol{u}_{m_1}, \dots; \boldsymbol{u}_1, \boldsymbol{u}_2, \boldsymbol{u}_3; \dots; \boldsymbol{u}_1, \boldsymbol{u}_2, \boldsymbol{u}_{m_1}; \dots; \dots.; \boldsymbol{u}_{m_1}, \boldsymbol{u}_{m_1-1}, \dots, \boldsymbol{u}_2$$

and find the corresponding optimal solutions

$$\overline{J_1}^{*}, \overline{J_2}^{*}, \overline{J_3}^{*}, \dots, \overline{J_{m_1}}^{*}; \overline{J_{12}}^{*}, \overline{J_{13}}^{*}, \dots, \overline{J_{1\,m_1}}^{*}; \overline{J_{21}}^{*}, \overline{J_{23}}^{*}, \dots, \overline{J_{2\,m_1}}^{*}; \\ \dots; \overline{J_{123}}^{*}, \dots, \overline{J_{12\,m_1}}^{*}; \dots; \overline{J_{m_1\,m_1-1\dots 2}}^{*}.$$

Regarding the minimization of the discretized OCP by CAM the set

$$\mathcal{J}^{*} = \{J_1^{*}; J_2^{*}, J_3^{*}, \dots, J_{m_1}^{*}; J_{12}^{*}, J_{13}^{*}, \dots, J_{1\,m_1}^{*}; J_{21}^{*}, J_{23}^{*}, \dots, J_{2\,m_1}^{*}; \dots; \\ J_{123}^{*}, \dots, J_{12\,m_1}^{*}; \dots; \dots; J_{m_1\,m_1-1\dots 2}^{*}; \overline{J_1}^{*}, \overline{J_2}^{*}, \overline{J_3}^{*}, \dots, \overline{J_{m_1}}^{*}; \\ \overline{J_{12}}^{*}, \overline{J_{13}}^{*}, \dots, \overline{J_{1\,m_1}}^{*}; \overline{J_{21}}^{*}, \overline{J_{23}}^{*}, \dots, \overline{J_{2\,m_1}}^{*}; \dots; \overline{J_{123}}^{*}, \dots, \overline{J_{12\,m_1}}^{*}; \dots; \\ \dots; \overline{J_{m_1\,m_1-1\dots 2}}^{*}\}.$$

will be obtained and the smallest elements of the set $\mathcal{J}^{*}$ will be the global optimum solution of the mentioned OCP based on the global property of CAM.

In some cases, in order to reduce the processing time of the algorithm based on OCP assumed conditions, it would be advantageous to take the subsequent suggestions.

- In some OCP, just use the combinations of the vectors being involved into the performance index.
- In some examples, just use state vectors or control vectors combinations to converge to the global optimum.

3. Generalization of Cutting Angle Method for Optimal Control Problems

Consider the following global optimization problem:

$$J(\boldsymbol{x},\boldsymbol{u},t) \to min$$
$$\text{subject to } \boldsymbol{x}(t),\boldsymbol{u}(t) \in \mathcal{S}^*,$$

where $\mathcal{S}^* = \{t \in \mathbb{R}^n_+ : \sum_i^r \boldsymbol{x}_i(t) = 1, \sum_i^r \boldsymbol{u}_i(t) = 1\}\}$ and r is the number of coordinates of the state and control vectors $\boldsymbol{x}(t),\boldsymbol{u}(t)$.

Let $G: \mathcal{S}^* \to \mathbb{R}$ be a positive Lipschitz functional defined on the unit simplex. Then it can be extended to a finite IPH functional $J(\boldsymbol{x},\boldsymbol{u},t)$ on the cone $\mathbb{R}^n_+$, which would coincide with $G(\boldsymbol{x},\boldsymbol{u},t) + \mathrm{c}$ on $\mathcal{S}^*$. i.e., $J(\boldsymbol{x},\boldsymbol{u},t) = G(\boldsymbol{x},\boldsymbol{u},t) + \mathrm{c}$ is an IPH functional on $\mathcal{S}^*$. Then, for each $\boldsymbol{x}(t),\boldsymbol{u}(t) \in \mathbb{R}^n_+$ define the support vectors

$$\begin{cases} L_{\boldsymbol{x}} = L = \left(\frac{J(x,u,t)}{x}\right) = \left(\frac{J(x,u,t)}{x_1}, \frac{J(x,u,t)}{x_2}, \dots, \frac{J(x,u,t)}{x_n}\right) \\ L_{\boldsymbol{u}} = \bar{\bar{L}} = \left(\frac{J(x,u,t)}{u}\right) = \left(\frac{J(x,u,t)}{u_1}, \frac{J(x,u,t)}{u_2}, \dots, \frac{J(x,u,t)}{u_n}\right) \end{cases}.$$

The n vectors $e^m = (0,\dots,0,1,0,\dots,0)$, with 1 in the mth position shall be used, and the corresponding support vectors $\begin{cases} L^m = \left(\frac{J(e^m,u,t)}{e^m}\right) \\ \bar{\bar{L}}^m = \left(\frac{J(x,e^m,t)}{e^m}\right) \end{cases}$, $m = 1,\dots,n$ shall be called basis vectors. A set of $k \geq n$ support vectors (and hence k known values of the functional $J(\boldsymbol{x},\boldsymbol{u},t)$ at k distinct vectors), $\begin{cases} \mathcal{K}_{\boldsymbol{x}} = \{L^k\}_{j=1}^k \\ \mathcal{K}_{\boldsymbol{u}} = \{\bar{\bar{L}}^k\}_{j=1}^k \end{cases}$. Let also the first n support vectors be the basis vectors. This choice of support vectors guarantees that all local (and hence global) minimizers of the auxiliary functionals $\begin{cases} H_k(\boldsymbol{x},\boldsymbol{u},t) = \max\limits_{j \geq k} \min\limits_{i=1,\dots,n} L_i^j \boldsymbol{x}_i \\ \bar{\bar{H}}_k(\boldsymbol{x},\boldsymbol{u},t) = \max\limits_{j \geq k} \min\limits_{i=1,\dots,n} \bar{\bar{L}}_i^j \boldsymbol{u}_i \end{cases}$ will be located in the interior of the unit simplex by algorithms. It always underestimates the value of

$$J(\boldsymbol{x},\boldsymbol{u},t): \begin{cases} H_k(\boldsymbol{x}) \leq J(\boldsymbol{x},\boldsymbol{u},t) \\ \bar{\bar{H}}_k(\boldsymbol{x}) \leq J(\boldsymbol{x},\boldsymbol{u},t) \end{cases}.$$

Hence, $\begin{cases} \mathscr{b}_k = \min_{x \in S^*} H_k(\boldsymbol{x}, \boldsymbol{u}, \boldsymbol{t}) \leq \min_{x \in S^*} J(\boldsymbol{x}, \boldsymbol{u}, \boldsymbol{t}) \\ \overline{\overline{\mathscr{b}}}_k = \min_{x \in S^*} \overline{\overline{H}}_k(\boldsymbol{x}, \boldsymbol{u}, \boldsymbol{t}) \leq \min_{u \in S^*} J(\boldsymbol{x}, \boldsymbol{u}, \boldsymbol{t}) \end{cases}$.

On the other hand, the sequence of its minima $\begin{cases} \{\mathscr{b}_k\}_{k=n}^{\infty} \\ \{\overline{\overline{\mathscr{b}}}_k\}_{k=n}^{\infty} \end{cases}$ is increasing, and then converges to the global minimum of $J(\boldsymbol{x}, \boldsymbol{u}, \boldsymbol{t})$ based on its function framework presented by [8], [5]. Furthermore, the following algorithms are presented for functional in general (or for OCP in particular) based on the CAM, which was formulated by [8], on function,

The CAM for OCP

Algorithm 1

Step (0). (Initialization)

a) Take points e^m, $m = 1, \dots, n$, and construct basic vectors $L^m = (\frac{J(e^m, u, t)}{e^m})$,
b) , $m = 1, \dots, n$.
c) Define the functional $H_n(x) = \max_{j \leq n} \min_{i=1,\dots,n} L_i^j x_i = \max_{j \leq n} L_j^j x_j$.
d) Set $k = n$.

Step 1. Find $x^* = \arg[\min_{x \in S^*} H_k(x)]$.

Step 2. Set $k = k + 1$ and $x^k = x^*$.

Step 3. Compute $L^j = \left(\frac{J(x^k, u, t)}{x^k}\right)$. Define the function

$$H_k(x) = \max_{j \leq k} \min_{i=1,\dots,n} L_i^j x_i = \max\{H_{k-1}(x), \min_{i=1,\dots,n} L_i^k x_i\}.$$

Go to Step 2.

Minimization of the auxiliary functional as the first step of algorithms 1 and 2 is considered as the most time consuming step. To solve this problem, the following theorem is proposed.

Theorem 1: *Let* $x^*(t) > 0$ (or $u^*(t) > 0$) *be a local minimizer of* $H_k(x,u,t)$ (or $\overline{\overline{H}}_k(x,u,t)$) *over the relative interior of* S^*, $riS^* = \{x(t) \in S^*, x(t) > 0\}$ (or $riS^* = \{u(t) \in S^*, u(t) > 0\}$). *Then there exists a subset* $\mathcal{L} = \{L^{j_1}, L^{j_2}, \dots, L^{j_n}\}$ (or $\overline{\overline{\mathcal{L}}} = \{\overline{\overline{L}}^{j_1}, \overline{\overline{L}}^{j_2}, \dots, \overline{\overline{L}}^{j_n}\}$) *of the set* $\mathcal{K}$ (or $\overline{\overline{\mathcal{K}}}$), *such that*

1) $x = \{d/L_1^{j_1}, d/L_2^{j_2}, \dots, d/L_n^{j_n}\}$ with $d = (\sum_i 1/L^{j_i})^{-1}$
   (or $u = \{d/\overline{\overline{L}}_1^{j_1}, d/\overline{\overline{L}}_2^{j_2}, \dots, d/\overline{\overline{L}}_n^{j_n}\}$ with $d = (\sum_i 1/\overline{\overline{L}}^{j_i})^{-1}$).
2) $\max\limits_{j \le k} \min\limits_{i=1,\dots,n} L_i^j / L_i^{j_i} = 1$
   (or $\max\limits_{j \le k} \min\limits_{i=1,\dots,n} \overline{\overline{L}}_i^j / \overline{\overline{L}}_i^{j_i} = 1$.
3) *Either* $\forall i: j_i = i$, *or* $\exists m: k_m > n, L_i^{j_m} > L_i^{j_i}, \forall i \neq m$
   (or *Either* $\forall i: j_i = i$, *or* $\exists m: k_m > n, \overline{\overline{L}}_i^{j_m} > \overline{\overline{L}}_i^{j_i}, \forall i \neq m$.).

Proof: Completely similar to the proof on function by [9].

The value of the auxiliary functional at x (or u) is $H_k(x,u,t) = d$ (or $\overline{\overline{H}}_k(x,u,t) = d$).

The following version of CAM for functionals works based on the above theorem, by examining all possible combinations of n support vectors.

Consider a set of k support vectors $\mathcal{K} = \{L^j\}_{j=1}^k$ (or $\overline{\overline{\mathcal{K}}} = \{\overline{\overline{L}}^j\}_{j=1}^k$), $L^j \in \mathbb{R}_+^n$ (or $\overline{\overline{L}}^j \in \mathbb{R}_+^n$). Let $\mathcal{J}$ denote $\{1,2,\dots,n\}$. From Theorem 1, the local minima of the auxiliary functional $H_k(x,u,t)$ (or $\overline{\overline{H}}_k(x,u,t)$ are combinations of n support vectors $\mathcal{L} = \{L^{j_1}, L^{j_2}, \dots, L^{j_n}\}$ (or $\overline{\overline{\mathcal{L}}} = \{\overline{\overline{L}}^{j_1}, \overline{\overline{L}}^{j_2}, \dots, \overline{\overline{L}}^{j_n}\}$) that satisfy the following conditions:

1) $\forall i, s \in \mathcal{J}, i \neq s: L_i^{j_i} < L_i^{j_s}$ (or $\overline{\overline{L}}_i^{j_i} < \overline{\overline{L}}_i^{j_s}$)
2) $\forall v \in \mathcal{K} \backslash \mathcal{L}, \exists i \in \mathcal{J}: L_i^{j_i} \ge v_i$ (or $\forall \overline{\overline{v}} \in \overline{\overline{\mathcal{K}}} \backslash \overline{\overline{\mathcal{L}}}, \exists i \in \mathcal{J}: \overline{\overline{L}}_i^{j_i} \ge \overline{\overline{v}}_i$).

The subset $\mathcal{L}$ (or $\overline{\overline{\mathcal{L}}}$), which satisfied conditions (I) and (II) above, is called a valid combination of support vectors.

Let $\mathcal{W}^k$ (or $\overline{\overline{\mathcal{W}}}^k$) denote the set of all valid combinations $\mathcal{L}$ (or $\overline{\overline{\mathcal{L}}}$) of k support vectors satisfying conditions (I) and (II):

$$\mathcal{W}^k = \{\mathcal{L} = \{L^{j_1}, L^{j_2}, \dots, L^{j_n}\}, L^{j_i} \in \mathcal{K}: (1) \forall i, s \in \mathcal{J}, i \neq s: \mathcal{L}_{ii} < \mathcal{L}_{si} \text{ and } (2) \forall v \in \mathcal{K} \backslash \mathcal{L}\ \exists i \in \mathcal{J}: \mathcal{L}_{ii} \ge v_i\}$$

(or $\overline{\overline{\mathcal{W}}}^{k} = \{\bar{\bar{\mathcal{L}}} = \{\bar{\bar{L}}^{j_1}, \bar{\bar{L}}^{j_2}, \ldots, \bar{\bar{L}}^{j_n}\}, \bar{\bar{L}}^{j_i} \in \bar{\bar{\mathcal{K}}}: (1) \forall i, s \in \mathcal{I}$,
$i \neq s: \bar{\bar{\mathcal{L}}}_{ii} < \bar{\bar{\mathcal{L}}}_{si}$ and $(2)\ \forall \bar{\bar{v}} \in \bar{\bar{\mathcal{K}}} \setminus \bar{\bar{\mathcal{L}}}\ \exists i \in \mathcal{I}: \bar{\bar{\mathcal{L}}}_{ii} \geq \bar{\bar{v}}_i\}$)

The problem of finding local minima of $H_k(x, u, t)$ (or $\bar{\bar{H}}_k(x, u, t)$ is translated into the problem of listing the elements of $\mathcal{W}^k$ (or $\bar{\bar{\mathcal{W}}}^k$). A simplistic approach is then:

Step 1. Construct all combinations $\mathcal{L}$ (or $\bar{\bar{\mathcal{L}}}$) satisfying (I)

Step 2. Check the obtained combinations against (II).

Then improve on this by taking into account the fact that all elements of $\mathcal{W}^k$ (or $\bar{\bar{\mathcal{W}}}^k$) not involving L^j (or $\bar{\bar{L}}^j$) (i.e., minima of $H_{k-1}(x, u, t)$ (or $\bar{\bar{H}}_{k-1}(x, u, t)$) do not need to be recomputed, hence, the algorithm takes the form:

Algorithm 2

Step (0). (Initialization)

a) Take points e^m, $m = 1, \ldots, n$, and construct basic vectors $L^m = \left(\frac{J(e^m, u, t)}{e^m}\right)$, $m = 1, \ldots, n$.
b) Define the function $H_n(x, u, t) = \max_{j \leq n} \min_{i=1,\ldots,n} L_i^j x_i = \max_{j \leq n} L_j^j x_j$.
c) Set $k = n$. Set $\mathcal{W}^k = \{\{L^1, L^2, \ldots, L^n\}\}$.
d) Calculate $d = (\sum_{i=1,\ldots n} 1/L_i^i)^{-1}$.

Step 1.

a) Retrieve all valid combinations $\mathcal{L}$ (i.e., $\mathcal{W}^k$).
b) Select $\mathcal{L} \in \mathcal{W}^k$.

Step 2.

a) $k = k + 1$.
b) Take $x^* = d/\text{diag}(\mathcal{L})$ and evaluate $J(x^*, u, t)$.
c) Compute $L^k = (J(x^*, u, t)/x^*)$. Define the function

$$H_k(x, u, t) = \max_{j \leq k} \min_{i=1,\ldots,n} L_i^j x_i = \max\{H_{k-1}(x), \min_{i=1,\ldots,n} L_i^k x_i\}$$

Step 3.

a) Check $\mathcal{W}^{k-1}$ against (II) and remove those that fail (II).
b) Move the remaining combinations into $\mathcal{W}^{k}$.

Step 4.

a) Construct all combinations $\mathcal{L}$ that involve L^{k} and satisfy (I).
b) Calculate $d = (\sum_{i=1,\ldots n} 1/L_i^{k_i})^{-1}$ for each such combination.
c) Add these combinations to $\mathcal{W}^{k}$.

Go to Step 1.

Remark 1: In Algorithm 2 if the diagonal of $\mathcal{L}$ (or $\bar{\bar{\mathcal{L}}}$) was dominated by L^{k}, this combination will be not moved into $\mathcal{W}^{k+1}$.

Remark 2: In Algorithm 2, each iteration the algorithm computes and checks the value of $J(x^*, u, t) - d$ (or $J(x, u^*, t) - d$) with the value of assumed stopping criteria. While it satisfies then the algorithm stopped.

It is obvious that the set $\mathcal{J}^*$ is countably finite. What is more, regarding the following proposition, $\mathcal{J}^*$ is not empty:

Theorem 2: *Suppose that the function f is continuous on the closed interval $[a, d]$. If $\mathfrak{m}$ and $\mathfrak{n}$ are, respectively, the absolute minimum and absolute maximum function values of f on $[a, d]$ so that* [10],

$$\mathfrak{m} \le f(x) \le \mathfrak{n} \text{ for } a \le x \le d$$

then

$$\mathfrak{m}(d-a) \le \int_a^d f(x)\,dx \le \mathfrak{n}(d-a).$$

Theorem 3 (The Mean-Value Theorem for Integrals): If f is continuous on the closed interval $[a, d]$, there exists a number c in $[a, d]$ such that [10],

$$\int_a^d f(x)\,dx = f(c)(d-a).$$

Proposition 1: The set of all optimal solutions $\mathcal{J}^*$ by CAM for discretized OCP is not empty.

Proof: the original OCP can be shown as follows

$$J(x,u,t) = \int_a^d E(x,u,t)\,dt$$
$$= \int_a^d F_1(x,u,t)\,dt + \int_a^d F_2(x,u,t)\,dt + \cdots$$
$$+ \int_a^d F_n(x,u,t)\,dt$$

where $E(x,u,t) = F_1(x,u,t) + F_2(x,u,t) + \cdots + F_n(x,u,t)$.

Based on Theorems 2 and 3, the following assertions shall be presented

$$\mathfrak{m}_1(d-a) \leq \int_a^d F_1(x,u,t)\,dt = F_1(c_1)(d-a)$$

$$\mathfrak{m}_2(d-a) \leq \int_a^d F_2(x,u,t)\,dt = F_2(c_2)(d-a)$$

$$\mathfrak{m}_n(d-a) \leq \int_a^d F_n(x,u,t)\,dt = F_n(c_n)(d-a) \qquad (5)$$

Next, (5) implies that

$$\mathfrak{m}_1(d-a) + \mathfrak{m}_2(d-a) + \cdots + \mathfrak{m}_n(d-a)$$
$$\leq \int_a^d F_1(x,u,t)\,dt + \int_a^d F_2(x,u,t)\,dt + \cdots$$
$$+ \int_a^d F_n(x,u,t)\,dt$$
$$= F_1(c_1)(d-a) + F_2(c_2)(d-a) + \cdots + F_n(c_n)(d-a)$$

$$\Rightarrow (\mathfrak{m}_1 + \mathfrak{m}_2 + \cdots + \mathfrak{m}_n)(d-a) \leq J(x,u,t) = c(d-a)$$

$$\Rightarrow \mathfrak{m}(d-a) \leq J(x,u,t) = c(d-a)$$

where $\mathfrak{m} = \mathfrak{m}_1 + \mathfrak{m}_2 + \cdots + \mathfrak{m}_n$ and $c = F_1(c_1) + F_2(c_2) + \cdots + F_n(c_n)$

Thus

$$\mathfrak{m} \leq c$$

Then there exists a $q \in \mathbb{R}$ such that $q = \mathfrak{m}$ or $q = c$ or $\mathfrak{m} < q < c$. Then implies that

$$J_q \leq J(x,u,t) \Rightarrow J_q \in \mathcal{J}.$$

where $q(d-a) \leq J_q = J(x_q, u_q, t)$ or $J_q = J(x_q, u_q, t) = q(d-a)$.

Thus $J_q \in \mathcal{J}^*$ and the set $\mathcal{J}^*$ is not empty ■.

Remark 3: Additionally this proposition shows that the global optimum of functional J belongs to the set $\mathcal{J}^*$ constantly.

Based on the Inheritance and Generalizability of function optimization algorithm to solve functional [7], it is represented that in Proposition 1and Remark 3 the globality of CAM to solve functions can be expanded in order to solve OCP in particular and functional in general. This procedure implies that the output of CAM for solving functional via UVCT is global. The validity of the theme will be demonstrated in the next examples.

4. Numerical Results

Example 1: The first minimization test OCP is

$$J = \int_0^1 (x^2(t) + u(t))dt$$

$$s.t\ \dot{x} = u(t)$$

$$x(0) = 0, x(1) = 0.5$$

For this problem, the interval $[0,1]$ was partitioned into 10 and 20 equal subintervals. So, for $n = 10: t_1 = 0, t_2 = 0.1, \dots, t_{10} = 1$ and for $n = 20: t_1 = 0, t_2 = 0.05, \dots, t_{20} = 1$. Then, CAM is applied for three iterations to minimize the mentioned OCP using both partitions. Then the results were compared which implies that the number of partitions is an impact factor for CAM, in finding the global optimal solution with stopping criterion$= 0.001$. The smallest element of $\mathcal{J}^*$ (optimum value of this example) is the minimum of J (see Figures 1 and 2). Thus

$$J^{*} = 0.0226$$

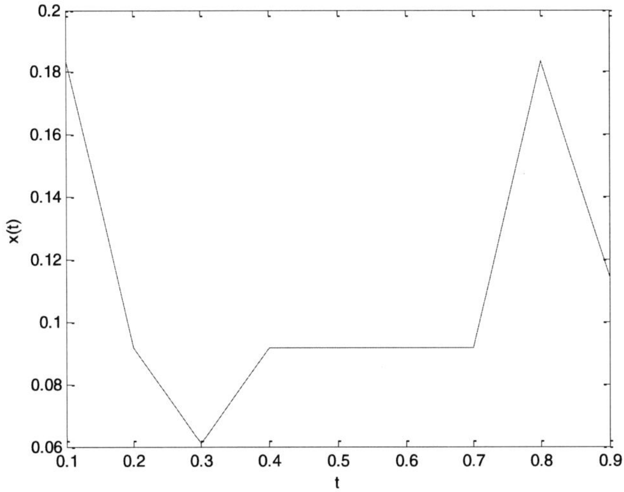

Figure 1. Optimum state vector of Example 1 for $n = 10$.

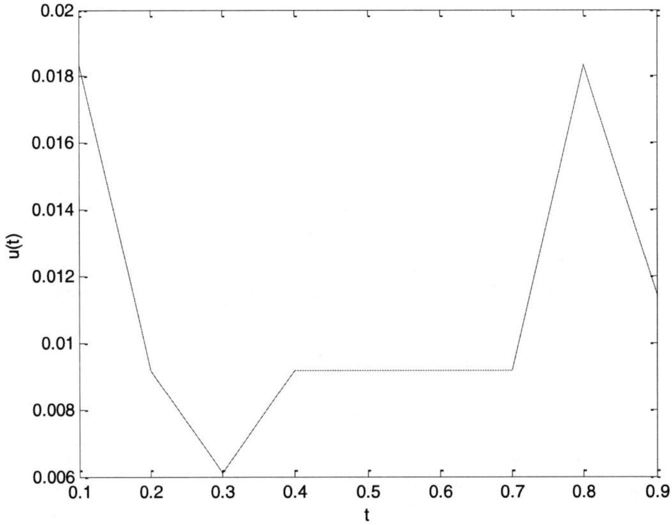

Figure 2. Optimum control vector of Example 1 for $n = 10$.

For $n = 20: t_1 = 0, t_2 = 0.05, \ldots, t_{20} = 1$, the step size $h = 0.05$ and discretized OCP is

$$J = 0.05 \sum_{i=1}^{19} (x^2(t) + u(t))$$

$$s.t\ x = 0.1 * u(t)$$

$$x(0) = 0, x(1) = 0.5,$$

Firstly, the CAM is used to find x^*,

After two iterations, the global minimizer is $J^* = 0.0056$. (see Figures 3 and 4).

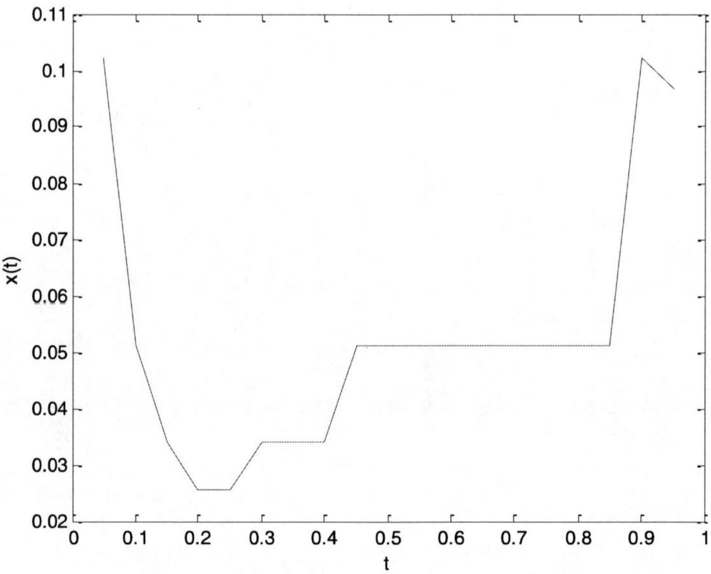

Figure 3. Optimum state vector of Example 1 for $n = 20$.

This example indicates that the solution of the problem will converge to global solution if appropriate number of partitions n increases.

Example 2: The second minimization test functional (Rubio, 1985) is

$$J = \int_0^1 (x_1^2(t) + x_2^2(t))dt$$

$$s.t\ \dot{x}_1 = x_2(t), \dot{x}_2 = 10x_1^3(t) + u(t)$$

$$x_1(0) = 0, x_1(1) = 0.1,$$

$$x_2(0) = 0, x_2(1) = 0.3.$$

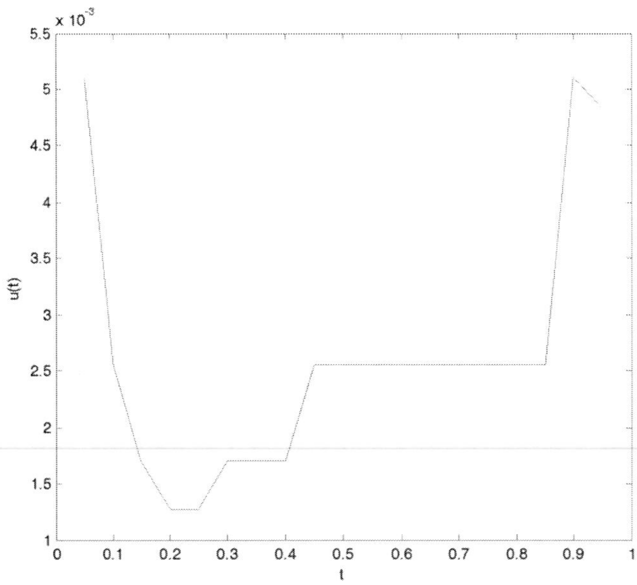

Figure 4. Optimum control vector of Example 1 for $n = 20$.

For this problem, the interval $[0, 1]$ was partitioned into 10 equal subintervals. So, for $n = 10$: $t_1 = 0, t_2 = 0.1, \dots, t_{10} = 1$. Then CAM was used in three iterations to minimize the mentioned OCP and Criterion= 0.001.

For this same example, the CAM is used to find $x_1{}^*$, $x_2{}^*$ based on UVCT (see Tables 1 and 2 and Figures 5 and 6). The global solution of the discretized OCP is 0.0218 by three iterations of CAM. The subsequent tables display that the CAM converges to the global solution with low number of iterations.

Table 1. Using CAM for three iterations and finding a local minimizer of Example 2 via UVCT

iteration	X1	J1
k=1	(0.1998, 0.0666, 0.0499, 0.0666, 0.0666, 0.0666, 0.0999, 0.1998, 0.1842)	2.1624
k=2	(0.2023, 0.0674, 0.0506, 0.0674, 0.0674, 0.0674, 0.1011, 0.2023, 0.1740)	2.1713
k=3	(0.2044, 0.0681, 0.0511, 0.0681, 0.0681, 0.0681, 0.1022, 0.2044, 0.1654)	2.1808

Table 2. Using CAM for three iterations and finding a global minimizer of Example 2 via UVCT

iteration	X2	J2
k=1	(0.1998, 0.0666, 0.0499, 0.0666, 0.0666, 0.0666, 0.0999, 0.1998, 0.1842)	0.0216
k=2	(0.2023, 0.0674, 0.0506, 0.0674, 0.0674, 0.0674, 0.1011, 0.2023, 0.1740)	0.0217
k=3	(0.2044, 0.0681, 0.0511, 0.0681, 0.0681, 0.0681, 0.1022, 0.2044, 0.1654)	0.0218

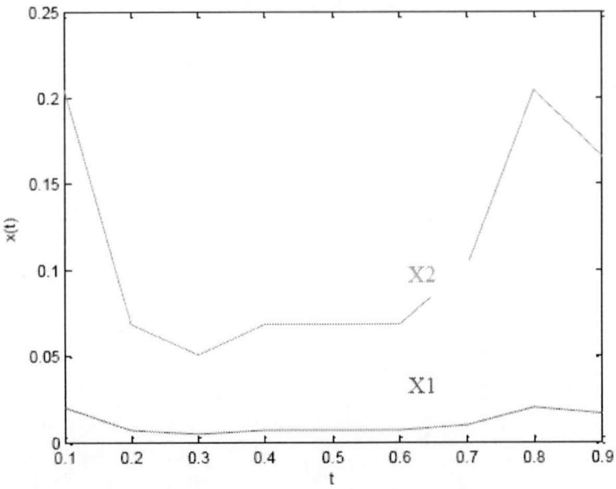

Figure 5. Optimum state vectors of Example 2 for $n = 10$.

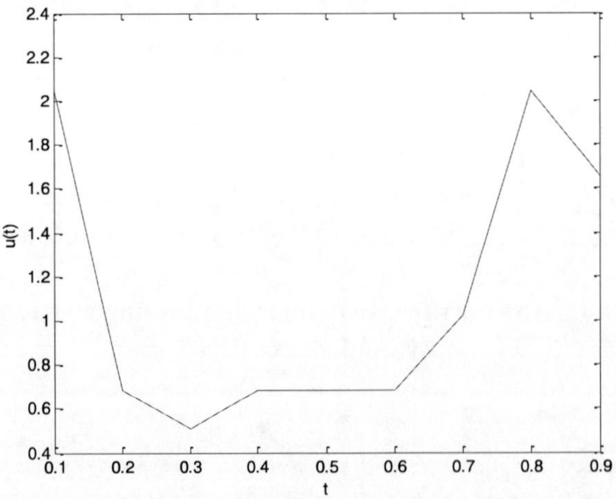

Figure 6. Optimum control vector of Example 2 for $n = 10$.

Example 3: Continuous Stirred Tank

The third minimization test functional (Kirk, 1970) is

$$J = \int_0^{0.78} (x_1^2(t) + x_2^2(t) + 0.1u^2(t))dt$$

$$s.t\ \dot{x}_1 = -(x_1(t) + 0.25) + (x_2(t) + 0.5)\exp\left(\frac{25x_1(t)}{x_1(t)+2}\right) - (1 + u(t))(x_1(t) + 0.25),$$

$$\dot{x}_2 = 0.5 - x_2(t) - (x_2(t) + 0.5)\exp\left(\frac{25x_1(t)}{x_1(t) + 2}\right)$$

$$x(0) = [0.05\ 0]^{\mathrm{T}}$$

For this problem, the interval $[0, 0.78]$ was partitioned into 10, 100, 200, 300 and 400 equal subintervals. So, for

$$n = 10\colon \mathfrak{h}_1 = 0, \mathfrak{h}_2 = 0.1, \dots, \mathfrak{h}_{10} = 0.78;$$
$$n = 100\colon \mathfrak{h}_1 = 0, \mathfrak{h}_2 = 0.01, \dots, \mathfrak{h}_{10} = 0.78;$$
$$n = 200\colon \mathfrak{h}_1 = 0, \mathfrak{h}_2 = 0.005, \dots, \mathfrak{h}_{10} = 0.78;$$
$$n = 300\colon \mathfrak{h}_1 = 0, \mathfrak{h}_2 = 0.0033, \dots, \mathfrak{h}_{10} = 0.78;$$
$$n = 400\colon \mathfrak{h}_1 = 0, \mathfrak{h}_2 = 0.0026, \dots, \mathfrak{h}_{10} = 0.78.$$

where $\mathfrak{h} = \frac{0.78-0}{n}$ is step size of discretized OCP. Consequently, the step size $\mathfrak{h}$ is equal to $0.1, 0.01, 0.005, 0.0033, 0.0026$ respectively. Then CAM used three iterations to minimize the mentioned OCP and stopping criterion= 0.001.

For this example, the CAM is used to find global optimum based on UVCT.

For $n = 10$:

$x_1^* =$
$(0.2041, 0.1020, 0.0680, 0.1020, 0.1020, 0.1020, 0.1020, 0.2041, 0.0137)$

$$x_2^* = (-0.0113, -0.0057, -0.0038, -0.0057, -0.0057, -0.0057, -0.0057, -0.0113, -0.0008)$$

$$J_{n=10}^* = 0.3163$$

for $n = 100$, $n = 200$, $n = 300$, $n = 400$, the solutions by CAM are $J_{n=100}^* = 0.0937$, $J_{n=200}^* = 0.0583$, $J_{n=300}^* = 0.0402$, $J_{n=400}^* = 0.0290$and step size $\mathfrak{h} = 0.01, 0.005, 0.0033, 0.0026$ respectively (see Figures 7, 8, 9, 10, 11, 12, 13, 14, 15 and 16). The global solution of this example by three iterations of CAM is 0.0290. As a comparison, this example was solved by 10 iterations of DISOPE algorithm and result was 0.028 (Rohanin, 2005). This example indicates that the CAM converges to the global solution faster than DISOPE algorithm numerically does.

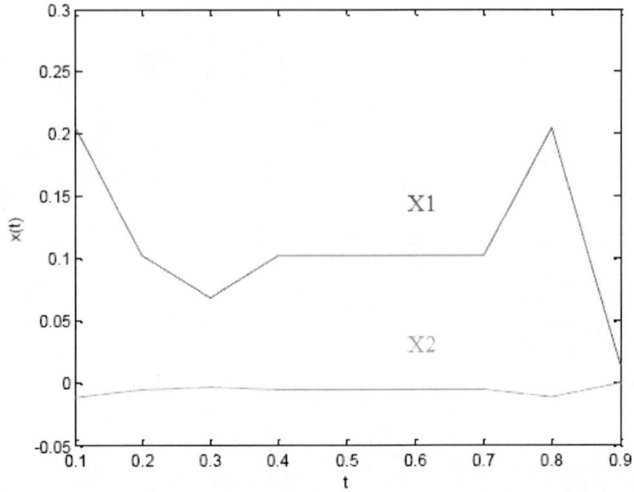

Figure 7. Optimum state vectors of Example 3 for $n = 10$.

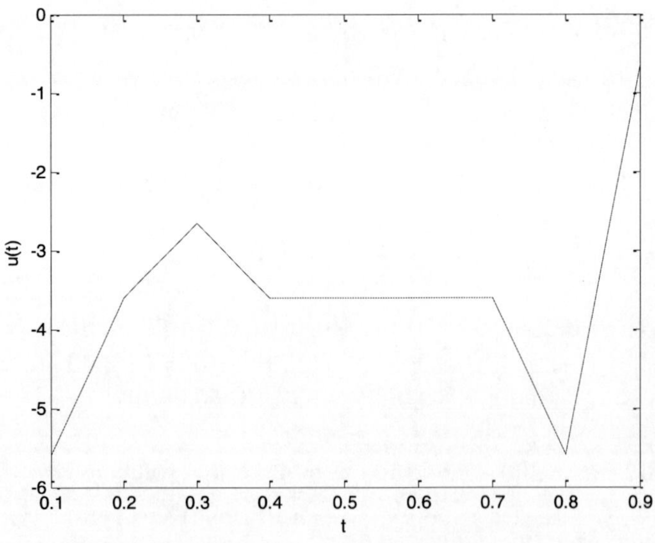

Figure 8. Optimum control vector of Example 3 for $n = 10$.

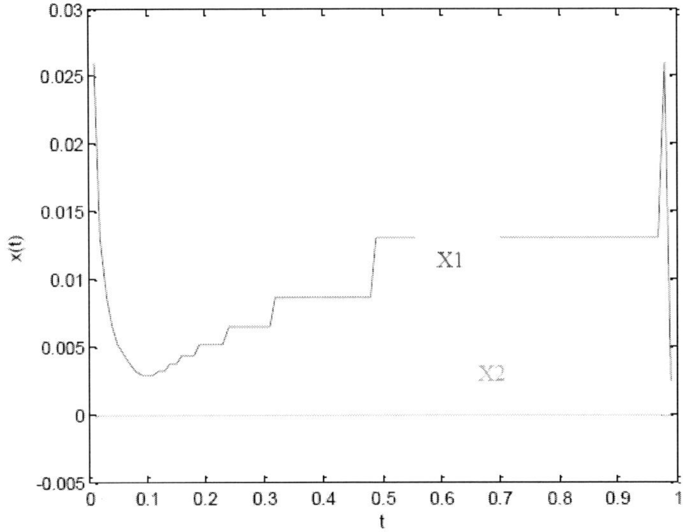

Figure 9. Optimum state vectors of Example 3 for $n = 100$.

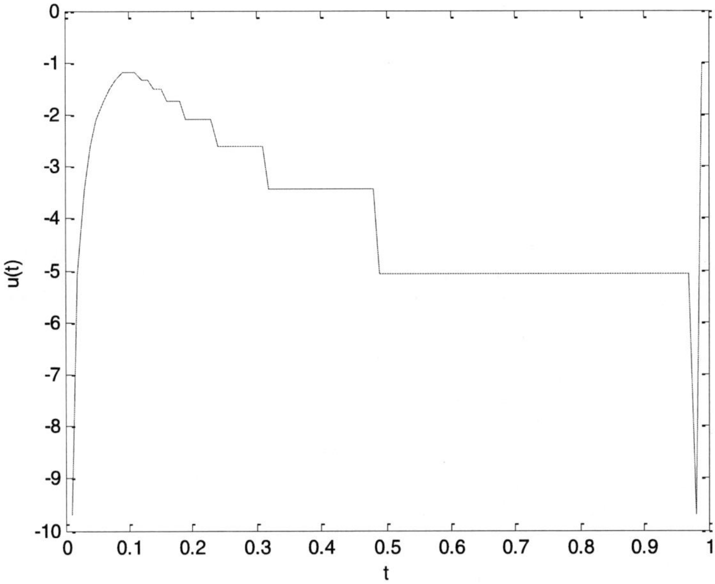

Figure 10. Optimum control vector of Example 3 for $n = 100$.

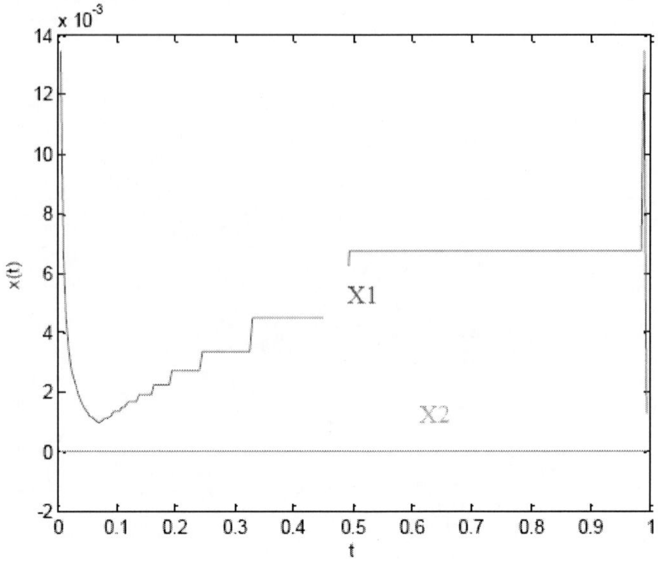

Figure 11. Optimum state vectors of Example 3 for $n = 200$.

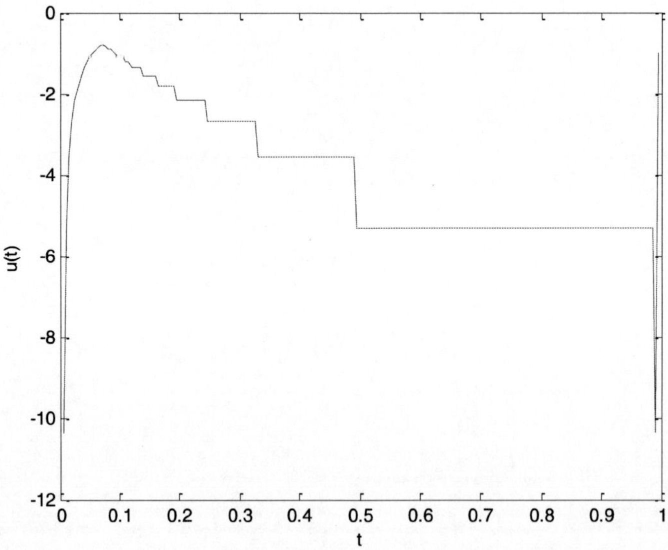

Figure 12. Optimum control vector of Example 3 for $n = 200$.

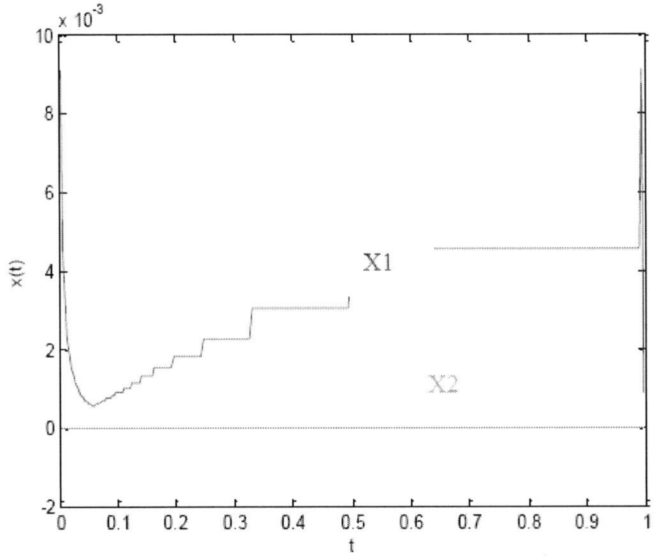

Figure 13. Optimum state vectors of Example 3 for $n = 300$.

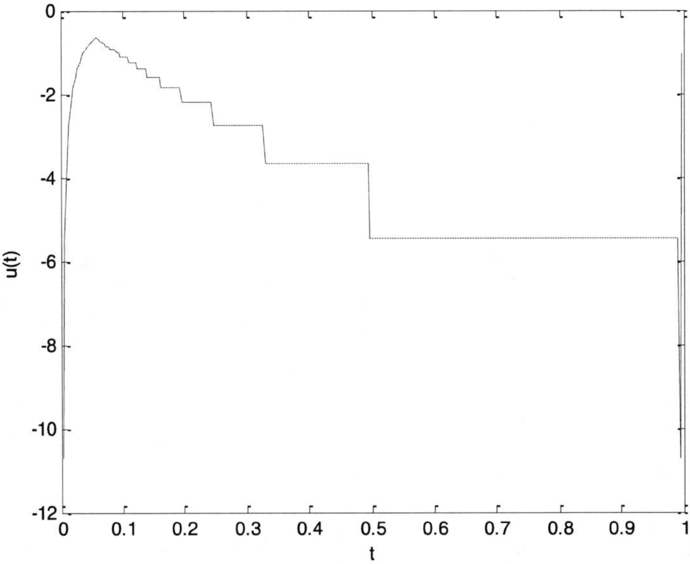

Figure 14. Optimum control vector of Example 3 for $n = 300$.

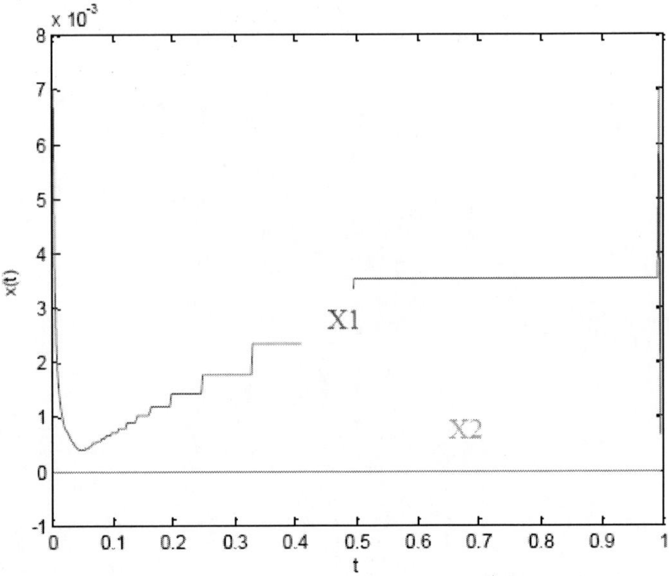

Figure 15. Optimum state vectors of Example 3 for $n = 400$.

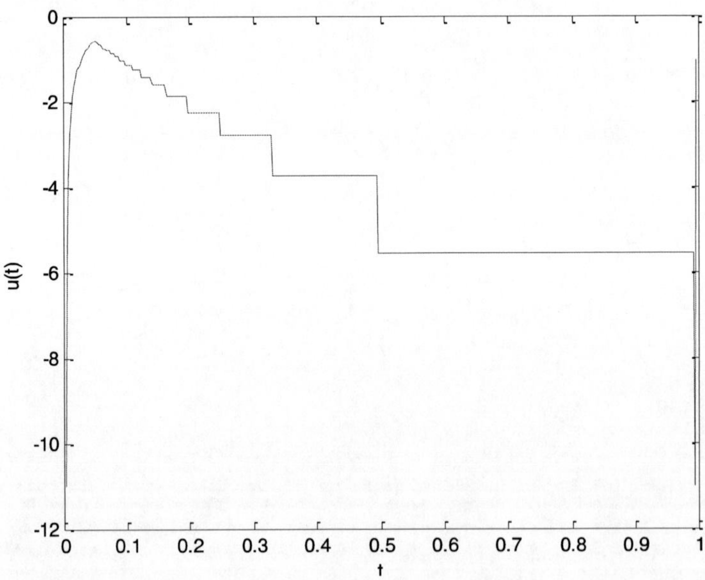

Figure 16. Optimum control vector of Example 3 for $n = 400$.

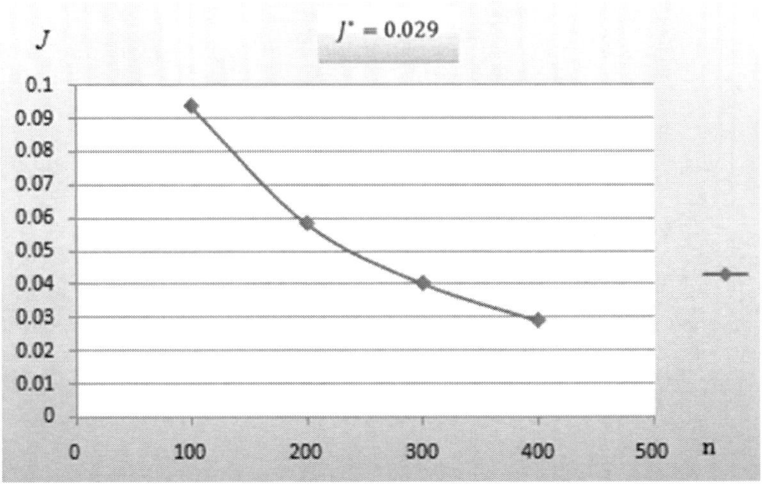

Figure 17. Convergence of the solution to global minimum with suitable n for Example 3.

This numerical finding implies that with a suitable number of partitions the CAM converges to a global solution .In addition, the main feature of this method is its convergence to global solutions in the lowest number of iterations (see Figure 17).

Example 4: The forth minimization test functional (Fard and Borzabadi, 2007) is

$$J = \int_0^1 (u^2(t))dt$$
$$s.t\ \dot{x}(t) = x^2(t) + u(t),$$
$$x(0) = 0, x(1) = 0.5,$$

For this problem, the interval $[0,1]$ was partitioned into 10 equal subintervals. So, for $n = 10: t_1 = 0, t_2 = 0.1, \ldots, t_{10} = 1$. Then CAM used two iterations to minimize the mentioned OCP and Criterion= 0.01 (The step size is $h = 0.1$). For this example, the CAM is used to find x^* and u^* based on UVCT.

$$x^* = (0.0002, 0.0001, 0.0000, 0.0001, 0.0001, 0.0001, 0.0001, 0.0002, 0.0002)$$
$$u^* = (0.1998, 0.0666, 0.0499, 0.0666, 0.0666, 0.0666, 0.0999, 0.1998, 0.1842)$$

The global solution of the discretized OCP is 0.0214 by two iterations of CAM (see Figures 18 and 19) and obtained result in the literature (Fard and Borzabadi, 2007) is 0.4447 by Genetic algorithm.

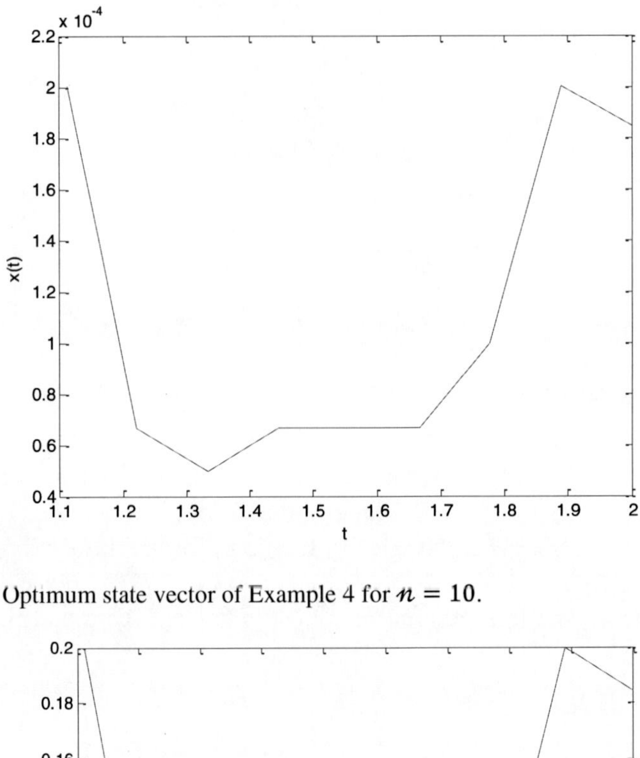

Figure 18. Optimum state vector of Example 4 for $n = 10$.

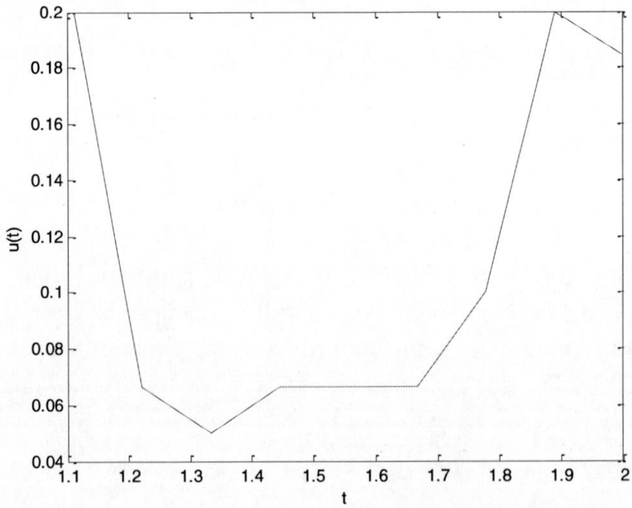

Figure 19. Optimum control vector of Example 4 for $n = 10$.

5. SUGGESTION

In order to study further the application of CAM to functionals for decreasing the time taken for convergence, the following suggestions are followed,

1) To decrease the number of vectors combinations in UVCT, it is more appropriate to substitute the vectors involved into the performance index. In particular, in some examples it is better to use only the state vectors (or control vectors) combinations to converge to the global optimum with reference to the constraints and the objective functionals.
2) Depending on the problems of forms it is better to find a suitable number of partitions experimentally in discretized OCP which is a major factor for rapidly converging to global solution.
3) In some examples, selecting the suitable stopping criterion shall help CAM to converge to best solution in OCP.

CONCLUSION

This paper proposed and investigated a direct CAM for optimal control problems via UVCT. The main finding of this study revealed that through the appropriate number of partitions, CAM can more practically converge to the best solution in the lowest number of iterations than any other algorithms in optimal control problems based on UVCT.

REFERENCES

[1] M. G. Singh, and A. B. Titli, *Systems: Decomposition, Optimization and Control*, Oxford: Pergamon Press, (1978).

[2] P. D. Roberts, An algorithm for optimal control of nonlinear systems with Model-Reality Differences, *12th IFAC World Congress on Automatic Control*, (1993).

[3] V. M. Becerra, *Development and Applications of Novel Optimal Control Algorithms*, Dept. of Electrical, Electronics and information Engineering, City University, UK: PhD thesis, (1994).

[4] Rohanin, *Theory and Implementation of Gradient based Modifications of a Model-Reality Differences Algorithm for Solving Nonlinear Optimal Control Problems*, Universiti Teknologi Malaysia: Ph.D thesis, (2005).

[5] A.M. Bagirov, and A.M. Rubinov, Global minimization of increasing Positively homogeneous functions over the unit Simplex, *Annals of Operations Research, Netherlands: Kluwer Academic Publications*, (2000).

[6] M. Andramonov, A. Rubinov, and B. Glover, Cutting Angle Methods in Global Optimization, *Applied Mathematics Letters 12, Elsevier Science Ltd*, (1999).

[7] S. Seyedi, A. Rohanin, and A. A. Mohd Ismail, Capability of the Function Optimization Algorithms for Solving Functionals, *ICMAE'10, August 2010, Kuala Lumpur, Malaysia.*

[8] A. M. Rubinov, *Abstract Convexity and Global Optimization*, Published by Kluwer Academic Publishers, (2000).

[9] A. M. Bagirov, and A.M. Rubinov, Cutting Angle Method and a Local Search. *Jurnal of Global Optimization, Netherlands: Kluwer Academic Publishers*, (2003).

[10] L. Lethold, *The Calculus 7*, Published by Harper Collins College Pablishers Inc: United States of America, (1996).

[11] J. E. Rubio, *control and optimization the linear treatment of nonlinear problems*, Manchester University Press, (1985).

[12] D. E. Kirk, *Optimal Control Theory: An Introduction*, London: Springer-Verlag, (1970).

[13] O. S. Fard, and A. H. Borzabadi, Optimal Control Problem, Quasi Assignment Problem and Genetic Algorithm, Published by *Proceedings of World Academy of Science, Engineering and Technology*, Volume 21, (2007).

About the Editor

Dr. Iraj Sadegh Amiri
Senior Lecturer
Photonics Research Centre (PRC)
University of Malaya (UM)
E2-02-03-Phase 1-Pantai Hillpark, Pantai Murni,
Kuala Lumpur, Malaysia
Email: irajsadeghamiri@gmail.com

INDEX

T

U

V